Structural Dynamics

Structural Dynamics

MARTIN WILLIAMS

University of Oxford, UK

CRC Press
Taylor & Francis Group
Boca Raton   London   New York

CRC Press is an imprint of the
Taylor & Francis Group, an **informa** business

A SPON PRESS BOOK

CRC Press
Taylor & Francis Group
6000 Broken Sound Parkway NW, Suite 300
Boca Raton, FL 33487-2742

© 2016 by Martin Williams
CRC Press is an imprint of Taylor & Francis Group, an Informa business

No claim to original U.S. Government works

Printed on acid-free paper
Version Date: 20160421

International Standard Book Number-13: 978-0-415-42732-6 (Paperback)

Library of Congress Cataloging-in-Publication Data

Names: Williams, Martin, 1962- author.
Title: Structural dynamics / author, Martin Williams.
Description: Boca Raton : Taylor & Francis, 2016. | Includes bibliographical references and index.
Identifiers: LCCN 2015042112 | ISBN 9780415427326 (pbk.)
Subjects: LCSH: Structural dynamics.
Classification: LCC TA654 .W55 2016 | DDC 624.1/71--dc23
LC record available at http://lccn.loc.gov/2015042112

Visit the Taylor & Francis Web site at
http://www.taylorandfrancis.com

and the CRC Press Web site at
http://www.crcpress.com

In memory of Kenneth Williams

Contents

Preface

This book aims to provide a concise and accessible introduction to the principles and underlying theory of structural dynamics. It is aimed principally at students of civil engineering but should also be of interest to other engineering disciplines with an interest in vibrations. It is based on my experience in teaching undergraduates in engineering science at Oxford University, and also on a series of short courses I have given over many years, aimed at civil engineering practitioners with little prior experience of dynamics.

Dynamic behaviour of structures has grown in prominence over recent decades and is now covered to some extent in most engineering courses. The importance of dynamic response to extreme environmental loads such as earthquakes has been recognised for some time, and is an essential topic for engineers working in parts of the world where these threats are prevalent. Increasingly, however, as more slender, daring structures are built, vibration problems are spreading to all parts of the world and to many different types of structures.

At one end of the spectrum, an engineer might be concerned with preventing collapse of a major suspension bridge due to aero-elastic instability in high winds. At the other, the issue might be minimising vibrations due to a single human footstep in highly motion-sensitive environments such as clean rooms. The range of possible load cases is too broad to discuss all of them here, but of course, they all share the same theoretical underpinning, which is the focus of this text.

We begin with an introductory chapter, Chapter 1, highlighting some key examples of dynamic behaviour, and introducing some of the language and mathematics of dynamics. Chapter 2 then introduces most of the important principles of structural dynamics by focusing on single-degree-of-freedom systems – idealisations of structures in which the motion is described by only one displacement variable. In Chapter 3, we look at how this basic theory can be extended to multi-degree-of-freedom systems. The methods introduced here are the basis of finite element dynamic analysis software routinely used in design offices. Chapter 4 looks at vibrations of continuous systems such as beams, where an explicit solution can often be obtained by solving the governing differential equations, without recourse to numerical methods. It also explains how to create approximate single-degree-of-freedom representations of continuous or distributed systems.

Many dynamic loads such as extreme wind and earthquake are likely to cause structural damage, normally in the form of localised yielding in structural members, causing a loss of stiffness. This is the subject of Chapter 5, which presents a variety of approaches to nonlinear dynamic analysis. Finally, in Chapter 6 the topic of Fourier analysis and random vibrations is introduced. This is a difficult topic and not one that has yet entered the civil engineering mainstream, but its importance is likely to grow in the near future.

While dynamics is unavoidably one of the more mathematically demanding branches of structural analysis, I have tried to keep the maths in the main text as simple and sparse as possible, with the focus on the underlying principles. The main mathematical techniques involved are briefly summarised in a set of appendices.

While I hope I have come up with some novel and helpful ways of presenting the concepts of structural dynamics, I make no claims to originality of the material presented, all of which has been in the public domain for some time. My aim here has been to present key concepts in a clear and concise way to introduce the topic. I have provided an extensive reading list for those wishing to know more, or delve deeper.

Of course, writing any book is a major undertaking that requires the support of numerous individuals and institutions. I am grateful to colleagues and students at the University of Oxford's Department of Engineering Science and at New College, Oxford, who for many years have provided such a stimulating working environment. Over the years, I have benefitted greatly from interactions with many outstanding engineers who have shaped my views on the subject. I thank them all, but wish to mention in particular Peter McFadden, Tony Blakeborough and Edmund Booth. Staff at Taylor & Francis have been patient and encouraging throughout the rather long gestation period of this work, and I am grateful in particular to Tony Moore and Amber Donley for their always calm and constructive input. Last but not least, the love and support of my wife EJ and son Rowan have been vital to the completion of this work.

Photos of the Tacoma Narrows Bridge are reproduced by permission of University of Washington Libraries, Special Collections, ref. WA21424. Other photos within the text are my own.

MSW

About the front cover

The image on the front cover is of the US Air Force Memorial, adjacent to Arlington National Cemetery just outside Washington, DC. It was designed by the architects Pei, Cobb, Freed and Partners, with structural engineering by Arup, and was opened by President George W. Bush in 2006.

The form of the structure was inspired by the *bomb-burst* manoeuvre performed by the US Air Force's Thunderbird F-16 jets during flight displays, in which several jets arc outward from (apparently) a single point, as though thrown apart by an explosion, leaving a set of curved smoke trails in the sky. These trails are echoed by the three curved, tapered spires, whose heights range from 61 to 82 m, with the tallest spire only 4.2 m wide at its base.

The Memorial was chosen for this book, both for its striking design, inspired by dynamics and movement, and because it has interesting and challenging vibrational characteristics. As might be expected for such long, slender structures, the spires have low fundamental frequencies of vibration – 1.0 Hz for the shortest tower, and just 0.57 Hz for the tallest. In addition, the towers have quite a sharp-cornered, triangular cross-section, rendering them vulnerable to a form of aero-elastic instability known as *galloping*, in which large and potentially damaging oscillations are created by the interaction between the structure and crosswinds. In most instances, galloping can be avoided by altering the shape of the structure to a more streamlined one, but in this case such a change would have destroyed the aesthetics of the Memorial.

The solution adopted was to increase the *damping* of the structure (that is, its inherent energy dissipation) to inhibit the galloping oscillations, using the novel approach of installing impact dampers within each spire. An impact damper comprises a heavy ball free to roll within a steel box. As the name implies, energy is dissipated through repeated inelastic impacts between the ball and the walls of the box. Impact dampers are not a particularly efficient form of energy dissipation and have not been widely used, but they were ideal in this case because they could be contained entirely within the towers, with no visual impact, and they are virtually maintenance-free.

In the USAF Memorial, each damper comprised a 0.5-m diameter, steel-coated lead sphere of mass 750 kg, contained in a box lined with a synthetic rubber chosen to maximise the energy absorption on impact. Different numbers of dampers were needed in each tower. An impact damper is most effective if it is located at a point in the structure that experiences relatively large motions, so that the rattling around of the ball is maximised. The tallest spire thus has six of these dampers (a total mass of 3 tonnes) positioned in a stack at around three-quarters of the way up – the cross-section above this point being too narrow to accommodate them. Damper performance was verified by lab tests and *in situ* monitoring of the completed structure, which showed that the levels of damping, and therefore the dynamic performance of the structure, were comfortably within design limits.

Symbols and abbreviations

Normal SI units are shown in brackets, although of course any consistent set of units can be used.

a	Acceleration (m/s^2)
$\mathbf{A}$	Amplification matrix
a_0	Rayleigh mass-proportional damping coefficient (s^{-1})
a_1	Rayleigh stiffness-proportional damping coefficient (s)
c	Damping coefficient (Ns/m)
$\mathbf{c}$	Damping matrix
$\mathbf{C}$	Generalised damping matrix
c_{crit}	Critical damping coefficient (Ns/m)
D	Dynamic amplification factor – the ratio of the peak displacement due to a dynamic load to that achieved under a static load of the same amplitude
E	Young's modulus (Pa)
$E(x)$	Expected, or mean, value of random variable x
$E(x^2)$	Mean square value of random variable x
e_{amp}	Amplitude error of a numerical solution method
$\mathbf{e}_i$	Error vector at step i
e_{per}	Period error of a numerical solution method
f	Frequency (Hz)
$\mathbf{f}$	Force vector
F	Force (N)
$\mathbf{F}$	Generalised force vector
$f(x)$	Probability density function for a random variable x
F_b	Base shear (N)
F_D	Damper force (N)
f_n	Natural frequency (Hz)
g	Acceleration due to gravity = 9.81 m/s^2
h	Height (m)
$H(s)$ or $H(\omega)$	Transfer function of a dynamic system
$h(t)$	Unit impulse response of a dynamic system
i	Mode number, or electrical current (A)
$\mathbf{i}$	Vector of influence coefficients in ground motion analysis
I	Second moment of area (m^4)
k	Stiffness (N/m)
$\mathbf{k}$	Stiffness matrix
$\mathbf{K}$	Generalised stiffness matrix
k^*	Generalised stiffness of equivalent SDOF system (N/m)

k_1	Stiffness of hysteretic component of bilinear stiffness model (N/m)
k_2	Post-yield stiffness in bilinear stiffness model (N/m)
L	Length (m)
L_i	Excitation factor for mode i
L_i/M_i	Modal participation factor for mode i
L_i^2/M_i	Participating mass for mode i (kg)
m	Mass (kg) or mass per unit length of beam (kg/m)
$\mathbf{m}$	Mass matrix
M	Bending moment (Nm)
$\mathbf{M}$	Generalised mass matrix
m^*	Generalised mass of equivalent SDOF system (kg)
M_i	Generalised, or modal, mass for mode i (kg)
p	Force per unit length (N/m)
$p(x)$	Probability that a random variable x takes a particular value
$P(x)$	Cumulative probability density function for a random variable x
p^*	Generalised loading of equivalent SDOF system (N)
P_i	Modal force for mode i of a continuous system (N)
R	Stiffness force (N)
$\mathbf{R}$	Vector of stiffness forces
$R_{xx,y}(\tau)$	Autocorrelation function of a variable $x(t)$
R_y	Yield force of non-linear spring (N)
s	Laplace variable
S	Shear force (N)
S_a	Spectral acceleration (m/s²)
S_d	Spectral displacement (m)
$\mathbf{s}_i$	Exact solution vector at step i
S_v	Spectral velocity (m/s)
$S_{xx}(\omega)$	Power spectral density of a variable $x(t)$
t	Time (s)
T	Period (s)
T_d	Damped natural period (s)
T_F	Period of harmonic force (s)
T_n	Natural period (s)
$\mathbf{u}$	Mode shape vector (eigenvector) of a lumped MDOF dynamic system
U	Total energy of a vibrating system (J)
$\mathbf{U}$	Matrix of mode shapes written as column vectors
$u(t)$	Unit step function
$u(z)$	Mode shape of a continuous dynamic system
v	Velocity (m/s)
W	Work (J)
$W_{xx}(\omega)$	Single-sided power spectral density $= 2S_{xx}(\omega)$
x	Displacement (m)
$\mathbf{x}$	Displacement vector
$\mathbf{X}$	Vector of generalised, or modal, displacements
$\bar{x}$	Time-average of x
$\dot{x}$	Velocity (m/s)
$\ddot{x}$	Acceleration (m/s², but sometimes expressed as a fraction of g, the acceleration due to gravity)
x_g	Ground displacement (m)

$\ddot{x}_g$	Ground acceleration (m/s^2)
x_p	Accumulated plastic displacement (m)
x_y	Yield displacement (m)
$\bar{x}(s)$	Laplace transform of $x(t)$
$X(\omega)$	Fourier transform of $x(t)$
y	Relative displacement between a mass and the ground, $y = x - x_g$ (m)
$\mathbf{y}$	Vector of relative displacements
$\mathbf{Y}$	Vector of generalised relative displacements
z	Position co-ordinate in beam vibrations (m)

GREEK SYMBOLS

α	Beam mode shape coefficient, or velocity exponent for non-linear damper
β	Newmark velocity parameter
γ	Newmark displacement parameter
δ	Logarithmic decrement
$\delta(t)$	Dirac delta function
Δt	Time step (s)
ϕ	Phase angle (rads)
κ	Curvature
λ	Eigenvalue of a matrix
Λ	Diagonal matrix of eigenvalues (squared natural frequencies)
μ	Mass ratio in 2DOF system, or coefficient of friction, or displacement ductility of yielding structure, or mean value of a random variable
ψ	Shape function used to reduce a continuous system to an equivalent SDOF system
σ	Standard deviation
σ^2	Variance
σ_{xy}	Covariance of random variables x and y
θ	Angle turned by rotating vector $= \omega t$
τ	Time shift variable used in convolution and autocorrelation (s)
ω	Circular frequency $= 2\pi f$ (rad/s)
ω_n	Circular natural frequency (rad/s)
ω_F	Circular frequency of harmonic force (rad/s)
Ω	Harmonic frequency ratio $= \omega_F/\omega_n$
ξ	Damping ratio $= c/c_{crit}$

ABBREVIATIONS

CDM	Central difference method
DFT	Discrete Fourier transform
DOF	Degree of freedom
EPP	Elastic-perfectly plastic
FFT	Fast Fourier transform
MDOF	Multi-degree-of-freedom
ODE	Ordinary differential equation
PDE	Partial differential equation

PDF	Probability density function
PGA	Peak ground acceleration
PSD	Power spectral density
SDOF	Single-degree-of-freedom
TMD	Tuned mass damper
ZPA	Zero-period acceleration

Chapter 1

Introduction to dynamic systems

dynamic, adj.: Of or pertaining to force producing motion: often opposed to static. [Adapted from French *dynamique* (Leibnitz, 1692), adapted from Greek δυναμικοσ (*dynamikos*) powerful.]

Oxford English Dictionary

1.1 WHAT IS DYNAMICS AND WHY IS IT IMPORTANT?

Dynamics is the science of how and why things move. Its origins can be traced back to the great 17th-century scientists and mathematicians such as Newton, Hooke and Leibnitz. This book is concerned particularly with the dynamics of structures, whose properties of mass and stiffness render them prone to oscillatory motions often referred to as vibrations.

Any physical object will be subjected to forces that vary over its lifetime, and will respond to those forces by moving. In many instances, the variations in force may be slow, or small in magnitude, and the corresponding motion is unimportant. In this book, we will focus on structural systems that are subjected to significant time-varying forces, such that their dynamic response to those forces needs to be taken into account by the engineers designing, maintaining or using them.

This chapter describes some of the key features of dynamic systems. After introducing the basics of oscillatory motion, we will review the structural properties that influence dynamic behaviour, consider how to idealise structures for dynamic analysis and introduce the characteristics of important dynamic loads. This chapter should enable you to:

- Understand why dynamics is important to structural engineers,
- Get to grips with the basic vocabulary of structural dynamics,
- Understand the nature of oscillatory motion, especially harmonic motion, and be able to describe it in mathematical form,
- Appreciate the properties of a structure (mass, stiffness and damping) that govern how it behaves when subjected to dynamic loads, and know how to calculate or estimate them,
- Be familiar with the principal forms of dynamic loads to which structures can be subjected.

To start, we will attempt to understand the importance of dynamics to engineers through a few brief case histories.

1.1.1 Why dynamics matters

This section will tend to focus on the negative—what can go wrong if dynamic behaviour is not properly understood and dealt with by designers. Learning the lessons from failures is one of the most valuable ways of improving knowledge and achieving better designs in the future.

1.1.1.1 The Tacoma Narrows Bridge

The Tacoma Narrows is probably the single most famous instance of structural collapse caused by dynamic response. Its cause, often wrongly explained in textbooks, was complex and is still not the subject of universal agreement.

The bridge was constructed in the late 1930s across part of Puget Sound, just south of Seattle in northwest United States, and opened to traffic in July 1940. By the standards of the time, it was a long suspension bridge, though not the longest. With a main span of 853 m, it was considerably shorter than the 1280-m Golden Gate Bridge in San Francisco, opened three years earlier. Even before it was opened, the bridge suffered from vibration problems. Vertical and/or twisting motions of the deck were easily excited by crosswinds, earning the bridge the nickname Galloping Gertie. Nevertheless, it remained open while solutions to the problem were sought.

On 7 November 1940, there was reportedly a steady crosswind of 64 km/h (about 18 m/s). This generated spectacular twisting oscillations of the deck, which continued for some time and ultimately grew to such an amplitude that the bridge was torn apart (Figure 1.1). Fortunately, the lengthy period of oscillation prior to collapse allowed evacuation of the bridge. As a result, the only fatality was Tubby, a dog whose owner could not coax him from his stranded car in time.

A common (but wrong) explanation of the collapse is that it was caused by vortex shedding. A vortex is a rotating, low-pressure flow that can be set up when a fluid such as air moves around a bluff body, that is, one that does not have a streamlined shape. On a bridge deck, vortices tend to be 'shed' from the downwind edge in an alternating pattern, creating an oscillating vertical force at that edge. If the rate of oscillation of this force is in tune with the natural vibration characteristics of the structure, then it can cause very large vertical and/or twisting motions of the structure due to the phenomenon of resonance. This is a credible mechanism, which bridge designers must consider, but for the Tacoma Narrows some simple calculations can show that the vortex shedding rate could not have caused resonance. Instead, it is widely believed that the cause of the oscillations was *aero-elastic flutter*, a complex interaction of aerodynamic, elastic and inertial forces.

What is universally agreed upon is that the behaviour was caused by the shape of the bridge deck. The deck was unusually narrow for such a major bridge, about 12 m wide, and was supported by plate girders only 2.4 m deep. This gave it both poor aerodynamics, the boxy profile causing maximum interaction with the wind, and very low torsional stiffness—a fatal combination. Compare, for instance, with the Golden Gate, whose 27-m wide deck is stiffened by huge 7.6-m deep steel trusses. The modern generation of suspension bridges such as the Humber Bridge near Hull, England have reverted to more slender decks, but these are heavily stiffened box girders with better torsional characteristics than the Tacoma Narrows, and are aerodynamically shaped to optimise their dynamic interaction with the wind.

Regardless of disagreements over the precise mechanism of its excitation, it is unarguable that the Tacoma collapse caused a huge surge in awareness of dynamics as an issue for civil engineering structures.

Figure 1.1 Twisting vibrations and collapse of the Tacoma Narrows Bridge, Seattle: sequence of stills from the film taken by Prof. J.B. Farquharson, 7 November 1940. (Courtesy of University of Washington Libraries, Special Collections, ref: WA 21424.)

1.1.1.2 The Northridge and Kobe Earthquakes

Earthquakes occur from time to time as the earth's tectonic plates shift around under the action of convection currents in the mantle—the mass of material between the earth's core and its outer layer, or lithosphere. Stress builds up at the plate interfaces, or faults, until a rupture occurs along the fault, with relative movement between the plates on either side. The sudden, large release of energy radiates outward as a ground vibration, its amplitude decaying with distance. Structures located close to the rupture are subjected to severe base shaking for a period of around 10–20 s, sometimes longer. This, in turn, sets the masses within the structure vibrating, generating large inertia forces that must be carried by the structural members. Seismic design aims either to resist these dynamic forces or, if they are too large for this to be feasible, to permit the structure to sustain damage short of collapse.

Two particularly significant events, from a structural engineering viewpoint, were the Northridge earthquake, which struck the Los Angeles region on 17 January 1994, and the Kobe (or more properly the Hyogo-ken Nanbu) earthquake that shook the Kobe-Osaka conurbation in southwest Japan exactly one year later, on 17 January 1995. Although large earthquakes occur every few years, and several since 1995 have resulted in far greater loss of life, Northridge and Kobe remain particularly significant events because the scale and severity of damage were quite unexpected. Beforehand, there was a widespread belief that wealthy, technologically advanced societies, particularly California and Japan, could build to restrict earthquake damage to tolerable levels, and were well prepared to cope with the aftermath of a major earthquake. These events were to tell a very different story.

The Northridge earthquake was, for California, a relatively modest one, with a Richter scale magnitude of 6.7 (Richter's scale is a rather strange, logarithmic one, in which an increase in magnitude of 1 represents an approximately 32-fold increase in the energy released). It caused surprisingly widespread damage, with numerous major bridge and multi-storey building collapses (some examples are shown in Figure 1.2), 57 fatalities and many more left injured or homeless. The financial consequences were in excess of US$20, making a serious impact on the US economy.

The earthquake-resistant design community was still coming to terms with this event when the Kobe earthquake struck the following year; this was an even bigger shock. Severe structural damage was very widespread, around 6000 people lost their lives and tens of thousands were left homeless in the chaotic aftermath, as the disaster relief agencies failed to cope. Economic losses have been estimated at US$150 billion, a figure which sent financial shock waves around the developed world.

As with Northridge, the numbers are all the more frightening because, while this was a large earthquake with a Richter scale magnitude of 7.2, it was certainly not Japan's 'next big one'. That was to occur on 11 March 2011—an enormous, magnitude 9.0 event centred

Figure 1.2 Earthquake effects on structures: severe damage to a reinforced concrete highway bridge, a reinforced concrete frame and a masonry structure, despite strengthening by wall ties. (Photos by the author, Northridge California, January 1994.)

70 km offshore from the northeastern province of Tohoku. This led to nearly 16,000 deaths and extensive structural damage, including to the nuclear power plant complex at Fukushima. However, in this case, the overwhelming majority of the damage and fatalities were caused not directly by the shaking, but by the huge tsunami, or tidal wave, that was triggered by the undersea quake. In structural dynamics terms, therefore, this event was less of a game-changer than Kobe.

In the aftermath of Northridge and Kobe, the field of earthquake engineering has been re-energised, particularly in the United States and Japan. Substantial resources have been put into improving design codes, developing new seismic-resistant structural systems and devices, and building bigger and better dynamic testing facilities.

Nevertheless, substantial risk remains. The consequences of, say, a magnitude 8.0 earthquake centred on Tokyo are almost unthinkable, both in terms of the devastation that would be suffered by the city and its inhabitants, and the significant damage to the world's financial system that would follow.

1.1.1.3 The London Millennium Bridge

One of a number of somewhat ill-fated Millennium projects, this new footbridge over the River Thames in London (Figure 1.3) took just a few hours' use to earn itself the unwanted nickname 'the wobbly bridge'. Huge crowds crossing the bridge on its opening day on 10 June 2000 experienced alarming horizontal sway vibrations, sufficient to make walking

Figure 1.3 The London Millennium Bridge: (a) general view, (b) viscous dampers installed diagonally between transverse arms, (c) viscous damper connected between diagonal braces beneath the deck and (d) tuned mass damper installed on cross-beam beneath the deck.

difficult. The bridge was closed for two years while the cause was investigated and an effective but costly retrofit was installed. The problem was solved but the nickname proved durable; indeed, the fault seems to have captured the public's imagination rather more than the bridge itself.

To engineers, the importance of the Millennium Bridge was that it brought to wider public attention the phenomenon of *dynamic human-structure interaction*—whereby the dynamic behaviour of the human body combines with that of a structure in quite complex and often counterintuitive ways. The importance of this interaction is now recognised for a range of structures: footbridges, office floors, gymnasiums and sports stadiums, to name a few.

In the case of the Millennium Bridge, the mechanism of walking produced small lateral oscillations of the walkers' body mass, which had the effect of putting energy into the structure, causing increasing lateral sway vibrations. Initially it was thought that this required the crowd to be walking in synchrony with the bridge's sway vibrations, but more recent research suggests that the effect can occur even if the pedestrians' walking rates are randomly distributed. In other circumstances, people can act as energy absorbers, reducing the motion of the structure.

The solution to the lateral vibration problem was a costly retrofit in which *viscous dampers* were installed under the bridge deck. These are devices to absorb energy, similar to the shock absorbers used in cars to smooth the ride on bumpy surfaces. They operate by forcing a viscous fluid through narrow orifices, causing the fluid to heat up as it undergoes shear deformations. In the Millennium Bridge, most of the viscous dampers were installed within cross-braces in such a way as to be subjected to the relative motions between points on the bridge deck 16 m apart. They are therefore activated when the bridge bends in the lateral direction. Some further energy absorption was provided by *tuned mass dampers*, in which the vibration of a small mass attached to the bridge takes energy away from the vibration of the main structure.

The movie *Harry Potter and the Half-Blood Prince* includes an enjoyable sequence in which the Millennium Bridge is destroyed in a manner strikingly similar to the Tacoma Narrows collapse. The filmmakers thus bring together these two landmark events in the history of structural dynamics, and present them to a much wider (and younger) audience than I can hope to achieve.

The above examples go some way toward illustrating the range of dynamics problems in structural engineering. They range from rare, extreme events carrying a grave risk of collapse and/or loss of life, through to serviceability problems, where vibrations may be uncomfortable or distressing but pose no immediate safety risk. However, of course, the underlying principles of dynamic behaviour are the same in all cases.

1.2 BASIC DEFINITIONS

Like most scientific disciplines, dynamics has a language of its own that can be off-putting to the newcomer. While we will try to avoid excessive jargon, it is essential to introduce a few of the basic terms that are used throughout the subject.

The study of dynamics can be broken down into the analysis of *free* and *forced* vibrations. In free vibration, the system is set in motion by some initial disturbance from its static equilibrium, and is then allowed to move free from any further external forces. The subsequent motion is then a function purely of the properties of the structure. In particular, the mass and stiffness of the system govern the rate at which it chooses to oscillate, usually known as the *natural frequency*.

Forced vibration occurs when the system is excited by a time-varying external force. If the frequency of the loading is close to the natural frequency of the system, then very large motions can occur due to *resonance*. In the study of any vibration problem, it is nearly always best to begin by determining the free vibration characteristics, which then serve as key structural properties that are used in the calculation of the forced vibration response.

In modelling structures, it is normal to reduce a distributed-parameter system to one characterised by the motion of a finite number of nodal points. Each motion variable included in the analysis is termed a *degree of freedom*. Thus, systems that can reasonably be approximated in terms of a single displacement variable are known as *single-degree-of-freedom* (SDOF) systems. Most of the key elements of structural dynamics will be developed for the SDOF case in Chapter 2, before extending the analysis to cover *multi-degree-of-freedom* (MDOF) systems in Chapter 3. The number of natural frequencies a structure possesses is equal to the number of its degrees of freedom. Thus, while an SDOF system possesses a unique natural frequency, an MDOF system will have many *modes* of vibration, each with its own natural frequency and each associated with a particular deformation shape known as the *mode shape*.

For most of this book, we will focus on the analysis of *linear* systems, in which materials are loaded within their elastic range so that restoring forces are proportional to deformations. For linear systems, the principle of superposition states that the response to more than one input equals the sum of the responses to each input applied individually. In Chapter 5, we will introduce the analysis of *nonlinear* systems in which yielding or other forms of nonlinearity mean that superposition can no longer be applied, and different analysis methods must therefore be used.

1.3 DESCRIBING OSCILLATORY MOTION

1.3.1 Time and frequency domains

Dynamics is the study of systems whose behaviour varies over time. We will therefore frequently work with loads, displacements or other parameters that can be expressed mathematically as functions of time, or whose variation with time can be plotted on a graph.

We can also represent time series in terms of their *frequency* content. Frequency is the number of oscillations that occur in a unit time, usually measured in cycles per second, or Hertz (Hz). A function that varies as the sine (or cosine) of time is known as *harmonic* and consists only of a single frequency. For example, Figure 1.4a shows a plot of the function

$$y = \sin(4\pi t)$$

The function completes two full cycles in each second, and so has a frequency of 2.0 Hz. We can express the variation with time as

$$y = \sin(2\pi f t) \qquad (1.1)$$

where the frequency f = 2.0 Hz. Alternatively, we can say that the *period* (i.e., the time to complete one full cycle) is T = 0.5 s; note that the period is simply the inverse of the frequency, and Equation 1.1 could have been written:

$$y = \sin\left(\frac{2\pi t}{T}\right) \qquad (1.2)$$

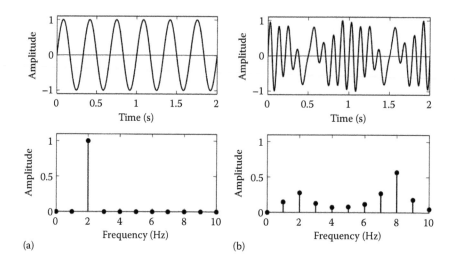

Figure 1.4 Two simple time series and their frequency spectra: (a) 2 Hz sine wave, (b) multi-sine.

Figure 1.4b shows a more complex time series, this time made up by summing numerous sinusoids having different frequencies f_i and amplitudes A_i. This can be expressed mathematically as

$$y = A_1 \sin(2\pi f_1 t) + A_2 \sin(2\pi f_2 t) + \ldots = \sum_{i=1}^{n} A_i \sin(2\pi f_i t) \qquad (1.3)$$

The simple sinusoid of Equation 1.1 is, of course, a special case of this formula with $n = 1$. If we plot the amplitude coefficients A_i as a function of f_i for each time series, we obtain a *spectrum*, showing how the amplitude of the signal is distributed between its various frequency components—these are shown below each time series in Figure 1.4.

Throughout this book, we will regularly return to the idea that a function can be represented in either the time or the frequency domain, the choice often governed by analytical convenience. The mathematics underlying the transformation between the two will be introduced later.

1.3.2 The basics of harmonic motion

In many instances the motion of structures is periodic, that is, it consists of regularly repeated, identical cycles of period T and frequency $f = 1/T$. Both of the time series in Figure 1.4 are periodic.

A particularly important form of periodic motion is *harmonic motion*. Many physical systems obey the laws of harmonic motion when allowed to vibrate freely, and under some forms of external loading. In a harmonic motion, the position x of an object varies sinusoidally with time t, according to:

$$x = X \sin \omega t \qquad (1.4)$$

The resulting variation of displacement with time is plotted in Figure 1.5. The parameter ω governs the rate of oscillations and is known as the circular frequency of the motion, normally measured in radians per second.

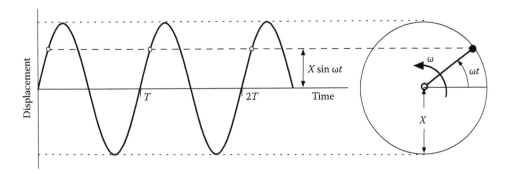

Figure 1.5 Harmonic motion and its equivalence to circular motion.

The expression *circular frequency* is apt because harmonic motion can usefully be visualised through its relationship to circular motion. Imagine you placed a small object on a horizontal turntable rotating with constant angular velocity ω. Viewed from above, the object repeatedly sweeps out circles, each full revolution of 2π radians taking a time $2\pi/\omega$. Next, squat down so that your eyes are level with the turntable, and look at the motion side-on. Disregarding any perspective effects, the object now appears to be moving from side to side. Its velocity appears greatest when it is at the middle of its travel, and it looks as though it momentarily comes to rest at the extreme left- and right-hand limits (when, in fact, it is moving either directly toward or away from you). In the time $2\pi/\omega$ it moves from one side to the other, then back to its starting position, that is, it completes one full cycle of harmonic motion.

The circular motion analogy is illustrated in Figure 1.5, where the harmonic oscillation corresponds to the vertical position of the particle at radius X rotating with constant angular velocity ω.

Given that one period of vibration equals the time taken for the rotating particle to trace out one complete circle, or 2π radians, it follows that the period T and frequency f are related to ω by

$$T = \frac{2\pi}{\omega}, \quad f = \frac{1}{T} = \frac{\omega}{2\pi} \tag{1.5}$$

and Equation 1.4 could also be written as

$$x = X \sin 2\pi f t = X \sin \frac{2\pi t}{T} \tag{1.6}$$

When plotting harmonic motion it is sometimes convenient to scale the time axis by the circular frequency ω, so that it has units of angle rather than time. This means that a single period of vibration plots as 2π radians. It also gives us a convenient way of introducing time lags and/or differences in initial conditions between signals having the same frequency. Imagine, for instance, two vibration signals x_1 and x_2, both with amplitude X and circular frequency ω, but while x_1 starts from zero and reaches its first peak after a quarter of a period, x_2 starts from a positive offset and reaches its peak some time τ earlier. We can express this mathematically as:

$$x_1 = X \sin \omega t, \quad x_2 = X \sin \omega (t + \tau)$$

or more conveniently as

$$x_1 = X \sin \omega t, \quad x_2 = X \sin(\omega t + \phi) \tag{1.7}$$

where $\phi = \tau/\omega$ is called the *phase angle* between x_1 and x_2. Here the plus sign implies that x_2 leads x_1 by the angle ϕ, whereas a minus sign would mean that x_2 lagged behind x_1 by the same amount. The meaning of the phase angle can be seen very clearly from Figure 1.6. With the time axis scaled by ω it is simply the horizontal separation between the curves. In terms of the analogy with circular motion, x_1 and x_2 both correspond to rotating vectors with angular velocity ω, but always separated by the angle ϕ.

Differentiating Equation 1.4 allows us to see how the velocity and acceleration of harmonic motion vary with time. Using the dot notation to represent differentiation with respect to time, the velocity $\dot{x}(= dx/dt)$ and acceleration $\ddot{x}$ $(= d^2x/dt^2)$ are

$$\dot{x} = \omega X \cos \omega t = \omega X \sin\left(\omega t + \frac{\pi}{2}\right) \tag{1.8}$$

$$\ddot{x} = -\omega^2 X \sin \omega t = \omega^2 X \sin(\omega t + \pi) \tag{1.9}$$

Looking at Equation 1.8, it is useful to note that a cosine wave is simply a sine wave offset by a phase angle of $\pi/2$. Therefore, when defining a harmonic function, it is very easy to switch between sine and cosine formulations, for instance, for analytical convenience or to suit the initial conditions.

Equations 1.4, 1.8 and 1.9 are plotted in Figure 1.7, the time axis again scaled by ω, so that it has units of angle rather than time. Some important relations between the displacement, velocity and acceleration can be seen. First, the amplitude coefficient increases by a factor of ω with each differentiation so that, for example, the velocity amplitude is ω times the displacement amplitude. Second, there is a phase angle of $\pi/2$ rads between the velocity and displacement, and π rads between the acceleration and displacement. This means, for example, that the velocity reaches its peak value at a time τ earlier than the displacement such that $\omega\tau = \pi/2$. The net effect of these phase differences is that the acceleration reaches its peak negative value at instants of maximum positive displacement, at which times the velocity is zero. The velocity takes its peak positive and negative values at times when both the displacement and the acceleration are zero.

Lastly, at any instant the acceleration and displacement are related by

$$\ddot{x} = -\omega^2 x \tag{1.10}$$

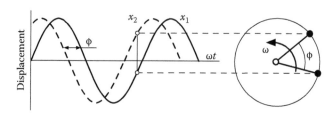

Figure 1.6 Two harmonic signals, differing only by a phase angle ϕ.

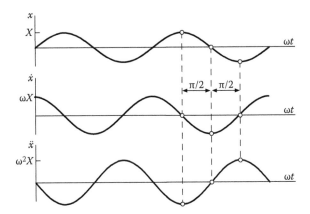

Figure 1.7 Relations between displacement, velocity and acceleration for a body undergoing harmonic motion.

Since Newton's second law of motion states that the acceleration of a body is proportional to the resultant force acting on it, Equation 1.10 implies that a body undergoing harmonic motion is always subjected to a force pulling it back toward its equilibrium position, with magnitude proportional to the distance from that position.

In some instances, it is mathematically more convenient to express harmonic motion using a complex exponential form, whose relation to the sinusoidal definition is given by Euler's equation:

$$z = Xe^{i\omega t} = X \cos \omega t + iX \sin \omega t \qquad (1.11)$$

where $i = \sqrt{(-1)}$. The quantity z is known as the complex sinusoid and, like the simple sinusoid of Equation 1.4 it satisfies the defining relationship for harmonic motion, $\ddot{z} = -\omega^2 z$. Plotting z on the Argand diagram (Figure 1.8a) again shows the relationship to circular motion. Harmonic motion is represented by a vector of magnitude X rotating at angular speed ω in the complex plane.

The sine or cosine form of the motion can be recovered (if necessary) by taking the real or imaginary part of z, or by combining with its complex conjugate

$$z^* = Xe^{-i\omega t} = X \cos \omega t - iX \sin \omega t, \qquad (1.12)$$

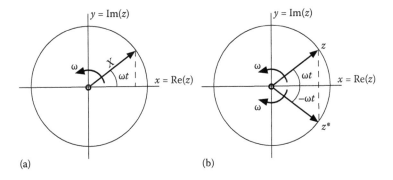

Figure 1.8 (a) Harmonic motion represented as a rotating complex vector and (b) recovering the real part as the sum of two counter-rotating complex vectors.

which represents an identical vector to z but rotating in the opposite direction (Figure 1.8b). We can then write

$$X \cos \omega t = \text{Re}(z) = \frac{1}{2}(z + z^*) = \frac{X}{2}(e^{i\omega t} + e^{-i\omega t}) \tag{1.13}$$

$$X \sin \omega t = \text{Im}(z) = \frac{-i}{2}(z - z^*) = -\frac{iX}{2}(e^{i\omega t} - e^{-i\omega t}) \tag{1.14}$$

Appendix A gives some advice on manipulating sine and cosine functions, and their equivalence to the complex exponential.

1.3.3 Phasors

The rotating vectors in Figures 1.5, 1.6 and 1.8 are known as *phasors*. By ignoring the fact that they are rotating and focusing instead on the phase angles between them, we can use phasors as an extremely easy way of analysing the relationships between different harmonic signals.

Suppose we have a harmonic motion with circular frequency ω and phase angle ϕ— remember that the phase angle just shifts the curve along the time axis without changing its shape, so that its value at time zero is $\sin \phi$ rather than zero. Using the complex exponential form, we can express this as

$$x = X \sin(\omega t + \phi) = \text{Im}\{Xe^{i(\omega t + \phi)}\} = \text{Im}\{Xe^{i\phi}.e^{i\omega t}\} \tag{1.15}$$

Here the first exponential term is the phasor, containing all the information about the amplitude and phase of the motion, and the second exponential tells us that x is oscillatory with frequency ω. Returning to the idea that harmonic motion can be visualised as a projection of the motion of an object on a turntable, imagine we now have two objects placed at different points. For each, the phasor is a line drawn from the centre of the turntable to the object, while the second exponential term describes the rotation of the turntable. As the turntable rotates, the two phasors both experience the same angular velocity, while their relative amplitudes and the phase angle between them stay the same.

The value of the phasor approach can be seen if we wish to compare or combine two signals having the same frequency but different amplitude and phase, for instance $a(t) =$

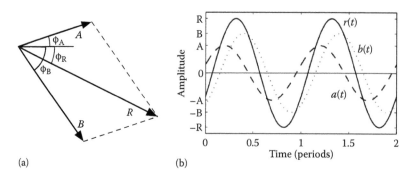

(a)   (b)

Figure 1.9 Combining two harmonic signals using phasors: (a) vector addition of the phasors for $a(t)$ and $b(t)$, (b) time histories of $a(t)$ and $b(t)$ and their resultant.

$A\sin(\omega t + \phi_A)$ and $b(t) = B\sin(\omega t + \phi_B)$. The amplitude and phase of these signals can be represented by the phasors shown in Figure 1.9a. The resultant $r(t)$ of the two signals can be found by simple vector addition of the two phasors as shown, giving a vector of amplitude R at phase angle ϕ_R. For each of a, b and r, the variation with time is then simply the vertical component of the phasor as it rotates at constant angular velocity ω. This is plotted in Figure 1.9b.

1.3.4 Some everyday examples of harmonic motion

1.3.4.1 Pendulum

Figure 1.10 shows a simple pendulum comprising a small mass on a string, pinned to a rigid support at the top. So long as its lateral displacement is small compared to the length of the string, then $\cos\theta \approx 1$ and the tension in the string remains approximately constant at mg as the pendulum swings. Noting that $\sin\theta = x/L$, the restoring force trying to pull the mass back to the centre is just the horizontal component of the tension in the string, mgx/L, and this in turn results in an acceleration toward the centre of gx/L. This satisfies the definition of harmonic motion in Equation 1.10, with the circular frequency of the oscillations given by $\omega = \sqrt{(g/L)}$. Thus, for small oscillations, the motion is harmonic. At larger amplitudes, the motion will still be oscillatory but the tension in the string will vary appreciably and so the restoring force will no longer be directly proportional to x.

1.3.4.2 Bobbing in water

Harmonic motion can also be seen when a floating object bobs up and down in water, such as a rubber duck in the bath (Figure 1.11). If the floating duck is given a small downward displacement x, and the cross-section A at the water surface is roughly constant, then the volume of water displaced is just Ax. Archimedes' principle tells us that the upward force exerted on the duck as a result of the displacement is equal to the weight of water displaced, which is just the unit weight of water γ_w (a constant) times the volume. Therefore, again, the restoring force is directly proportional to the displacement, resulting in harmonic motion.

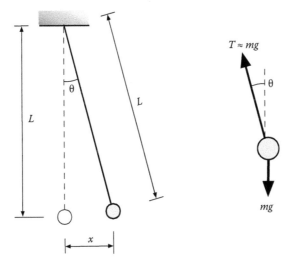

Figure 1.10 A simple pendulum: dimensions and free body diagram.

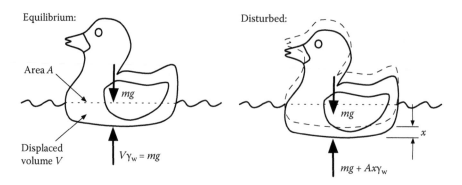

Figure 1.11 Vertical bobbing of a duck in water.

1.4 DYNAMIC PROPERTIES OF PHYSICAL SYSTEMS

The key structural properties needed for a dynamic analysis are the distributions of mass, stiffness and damping within the structure. In this section, we will consider why these are important and how they can be represented in a way that makes the analysis as straightforward as possible.

1.4.1 Mass and stiffness

The fundamental principle governing dynamics is Newton's second law of motion, which states that the resultant force on a body equals the rate of change of its momentum.* For systems with constant mass, this is normally written as

$$F = ma \tag{1.16}$$

where F is the resultant of the forces acting on the body under consideration, m is its mass and a its acceleration. Equation 1.16 tells us that motion occurs due to a lack of equilibrium, and that the key physical parameter governing the dynamic response is the mass. Mass is the fundamental measure of the quantity of a material, with SI units of kilograms (kg). It should not be confused with weight, which is the force exerted on a mass by a gravitational field. Force is measured in Newtons (N), where 1 N is the force required to accelerate a mass of 1 kg by 1 m/s². Since the acceleration due to gravity is $g = 9.81$ m/s², the weight of 1 kg is 9.81 N.

If a mass is restrained by an element having finite stiffness, such as a spring, then it has the potential to undergo oscillatory motion. Defining the origin of the motion as the position at which the spring is unstrained, then any motion x away from the origin will generate a restoring force R, as shown in Figure 1.12. In general, R may be any function of the spring extension x. If the spring is linear, then

$$R = kx \tag{1.17}$$

---

* From Sir Isaac Newton's *Philosiphae Naturalis Principia Mathematica* (1687): 'Lex II: Mutationem motus proportionalem esse vi motrici impressae, et fieri secundum lineam rectam qua vis illa imprimitur'. ('Law II: The alteration of motion is ever proportional to the motive force impressed, and is made in the direction of the right line in which that force is impressed'.)

Figure 1.12 Displacements and forces in a simple mass-spring system.

where the constant k is the stiffness, that is, the force required to give a unit deflection, with units of N/m. Note that R always opposes the motion, tending to pull the mass back toward the origin.

Consider, for example, the effect of a movement of the mass to the right of its equilibrium position. The spring force pulls it back to the left, causing it to decelerate. Eventually, the mass comes to rest and then starts moving back toward the origin. When it reaches the origin, the spring has returned to its unstrained length and so imposes no force. However, the mass now has a velocity, and hence a momentum, to the left and so continues to move past the origin. The spring now generates a restoring force to the right, and the above sequence repeats itself in the opposite direction. The mass thus oscillates to and fro, with a continuous exchange between force in the spring and momentum of the mass.

The behaviour can also be visualised in terms of conservation of energy. As the mass passes through the origin, its kinetic energy takes its maximum value. As the mass displaces from the origin, it slows down and the spring deforms, so that the kinetic energy reduces and the energy thus released is transferred to the spring in the form of elastic strain energy. At the limit of its travel, all the energy is stored in the spring and the kinetic energy is zero. The mass then begins to move back toward the origin and the energy transfer is reversed.

1.4.2 Damping

An idealised mass-spring system, once set in motion, will continue to oscillate indefinitely. In reality, of course, we know that a vibration will die away over time unless there is some external exciting force to keep it going. The reason the motion dies away is the presence of a variety of internal mechanisms, which take energy out of the system. This energy dissipation is known as *damping*. There are many possible forms of damping, for example:

- Energy is lost through friction between sliding surfaces (often known as dry damping, or Coulomb damping).
- Viscous damping describes the energy lost in lubricating fluids between moving components, or when fluid flows through an orifice, as in vehicle shock absorbers.
- Hysteretic damping (sometimes called structural damping) occurs through random vibration of the crystal lattice of a solid when it is subjected to repeated cycles of deformation.
- Radiation damping occurs when energy is transmitted from the system under consideration into some other medium and not returned, for example, from a building into the ground.
- Aerodynamic damping occurs due to friction between a vibrating body and the surrounding air.
- Similarly, hydrodynamic damping arises from the energy loss as a vibrating body moves through a fluid (e.g., an oil rig leg through seawater).

In any physical system, the energy dissipation will probably be due to a combination of several of these forms of damping. Modelling them explicitly is generally difficult, and introduces unhelpful mathematical complexity. Therefore, it is common simply to represent damping by a viscous dashpot, in which a piston pushes a viscous fluid such as oil through a narrow orifice. This results in the most mathematically simple formulation, and is a reasonable approximation to a variety of damping mechanisms, so long as the overall level of damping is reasonably small. This approach is not always appropriate, for example in systems where there is substantial friction. However, we will assume it is an acceptable approximation throughout the majority of this text.

The dashpot is represented diagramatically as shown in Figure 1.13, and produces a retarding force proportional to the change in velocity v across it:

$$F_D = cv \tag{1.18}$$

With force measured in N and velocity in m/s, the damping coefficient c has units of Ns/m. Note that the force on the mass caused by the dashpot is always in the opposite direction to the velocity.

If we plot damping force against velocity, then obviously we get a simple straight line of gradient c. A more interesting graph can be obtained by plotting the force against deflection. For a harmonic vibration we saw in Equations 1.4 and 1.8 that

$$x = X \sin \omega t; \quad \dot{x} = \omega X \cos \omega t$$

Using Equation 1.18, the damping force is

$$F_D = c\dot{x} = c\omega X \cos \omega t \tag{1.19}$$

Then substituting into the well-know trigonometric identity

$$\sin^2 \omega t + \cos^2 \omega t = 1$$

gives

$$\left(\frac{x}{X}\right)^2 + \left(\frac{F_D}{c\omega X}\right)^2 = 1 \tag{1.20}$$

This is the equation of the ellipse plotted in Figure 1.14, with one cycle of harmonic motion represented by one traverse around the ellipse. The area enclosed within the ellipse (integral of force with respect to distance moved) represents the work done by the damper

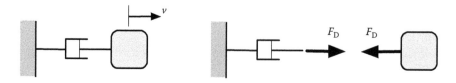

Figure 1.13 Velocity and forces in a simple mass-dashpot system.

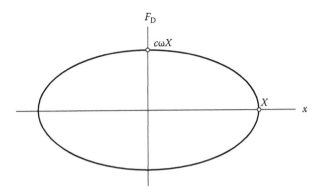

Figure 1.14 Force-displacement loop for a viscous damper undergoing a single vibration cycle of amplitude X.

force over the cycle, that is, the energy dissipation. Since the area of an ellipse is π times the product of the axis intercepts, the work W_D is

$$W_D = \pi c\omega X^2 \tag{1.21}$$

Or, put the other way round, if we were able to measure the energy dissipated per cycle of vibration, we could use this to estimate the viscous damping coefficient:

$$c = \frac{W_D}{\pi\omega X^2} \tag{1.22}$$

Since real damping is not often purely viscous, the actual force-displacement relationship may differ from the ideal ellipse. Even so, using Equation 1.22, it would be possible calculate an equivalent viscous damping coefficient that would give the same energy absorption for the same peak displacement. This gives a simple method of approximating the real damping mechanism in a mathematically convenient way.

Since it is (in most cases) only an approximate representation of reality, it can be rather difficult to assign an appropriate value to the damping coefficient. Unlike mass and stiffness, damping properties cannot easily be derived analytically from the physical properties of the system under consideration, and the kind of experimental data needed to apply Equation 1.22 will often be unobtainable. We will see that it is often more convenient to use a dimensionless parameter, the *damping ratio* ξ,

$$\xi = \frac{c}{2\sqrt{km}} \tag{1.23}$$

ξ is useful because it tends to take a similar value for structures of a similar type and material, and it is common simply to assume a ballpark value for analysis purposes, based on experience. For example, for civil engineering systems, the value of ξ is often taken to be in the range 0.02 to 0.05, though higher or lower values can occur.

1.4.3 Modelling physical systems for dynamic analysis

To perform an accurate dynamic analysis of a structure, the three key properties of mass, stiffness and damping must be realistically described. As stated above, it is common to represent damping by a simple, global parameter, and this is generally sufficient so long as the overall level of damping is small.

Mass and stiffness can usually be estimated more accurately. In simple cases, such as the vibration of a uniform beam, it is possible to create a continuum model in which the distribution of mass and stiffness are represented exactly by mathematical functions; this approach is explored in Chapter 4. For more complex systems, it is normal to represent them in discrete form, as assemblages of point masses connected by springs, with the displacements of the masses as the primary variables in the analysis. A key task for the analyst is choosing an appropriate level of model complexity.

Consider, for example, the flat-roofed portal frame shown in Figure 1.15. If the structure were subjected to a dynamic horizontal load, then at any instant it would be expected to adopt the deformed shape shown by the dashed line. Its distributed mass and stiffness can be discretized in a variety of ways, depending on the output required. In most analyses, the response variable of greatest interest would be the lateral deflection of the cap beam, so it would be possible to represent the structure by the simple single-degree-of-freedom mass-spring model shown. Here the spring represents the lateral sway stiffness of the frame, that is, the force required at cap beam level to produce a corresponding unit deflection; this can be obtained from a static structural analysis. The mass associated with the lateral sway degree of freedom is hard to determine precisely but can reasonably be approximated by lumping the masses of the various elements at their ends. So the mass in our SDOF model would represent the beam mass plus half of the column masses (the other half being lumped at their bases).

In some circumstances, more detailed output might be sought, for example vertical as well as horizontal motions, joint rotations or the deflections of intermediate points within members. The structure might then be discretized more finely, with the masses lumped at several points, and each section of beam or column between the mass points treated as a separate spring element. Such an approach can result in a rapid increase in model complexity, since each mass now possesses two degrees of freedom (horizontal and vertical deflections). As in any engineering modelling exercise, there is a need to balance the requirements of accuracy and completeness of the output against the desirability of a simple model that captures the key features of the behaviour with clarity.

These arguments can be extended to more complex structures. Figure 1.16 shows a three-storey frame; for simplicity, this is idealised as a 'shear-type' building, in which the floors are assumed rigid, so that the lateral deformation involves a shearing displacement between them. An obvious modelling approach would be to treat the lateral displacement of each storey as a degree of freedom, so that the structure could be modelled by the 3DOF mass-spring system shown, with each spring representing the sway stiffness of a storey and each

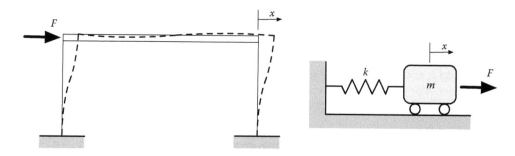

Figure 1.15 Single-storey portal frame with a lateral dynamic load, and its representation by a SDOF mass-spring system.

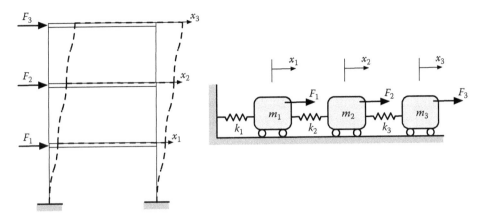

Figure 1.16 Idealised 'shear-type' three-storey frame structure, and its 3-degree-of-freedom mass-spring model.

mass representing the floor mass plus some fraction of the mass of the adjacent columns. However, more complex or simpler options might also be appropriate.

For hand analysis, it naturally makes sense to model a structural system using the fewest degrees of freedom that can realistically describe its behaviour; otherwise, the analysis quickly becomes unwieldy. If modelling the structure using a finite element program, it is easy for the analyst to discretize the structure to the desired level, though it is still wise to avoid unnecessarily complex models. While the rapid advances in computing power mean that large run times are much less of an issue than in years past, model complexity can still cause great difficulty in interpreting results.

1.5 DYNAMIC LOADS

The other key component of a dynamic system is the external loading, which causes the structure to move. Any externally applied loading or displacement will cause movement, but the more interesting and important cases generally involve loads that vary quite rapidly over time, and particularly those that oscillate at a similar frequency to the natural frequency of the structure, causing resonance. Dynamic loads can be classified into several types:

- *Impulsive*: The load is applied and/or removed suddenly, so that the structure is given a jolt. Examples include explosive blast or a single human jump.
- *Oscillatory, periodic*: The load cycles back and forth in a regular, repeating sequence. Such loading may be produced by rotating machinery or by simple human activities such as walking or repeated jumping on a floor.
- *Oscillatory, random*: The load fluctuates back and forth but in an irregular way, without obvious repetition, that cannot be neatly described by a simple mathematical function. Environmental loads such as wind, wave and earthquake fall into this category.

1.5.1 Impulsive loads

Figure 1.17 shows some examples of impulsive loads. In Figure 1.17a a large magnitude load is applied for a short duration dt, giving an impulse $I = F.dt$. If dt is very short, there is little time for the structure to move in response to the load. The effect of the impulse is therefore

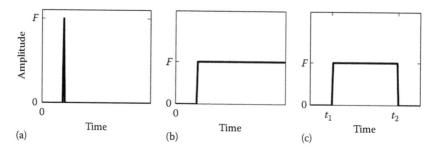

Figure 1.17 Examples of impulsive loads: (a) short-duration pulse, (b) step load: suddenly-applied and maintained indefinitely, (c) finite-duration step load: suddenly applied then suddenly removed.

to initiate motion of the structure (since the impulse causes a change in momentum), after which it undergoes free vibration. Usually the deformation of the structure continues to increase after the loading has finished, due to the momentum imparted to it.

Figure 1.17b shows a step load, in which a force is suddenly applied and then maintained. This can be thought of as a combination of an impulsive load (setting the system in motion, as above) and a static load, causing the structure to shift to a new equilibrium position.

Other types of impulsive load are possible and can be visualised as combinations of impulses or steps. For instance, the short-duration load in Figure 1.17c can be regarded as the sum of a positive step load at time t_1 and a negative one at time t_2.

1.5.2 Periodic loads

A periodic load is one that repeats itself regularly over time. If the variation with time is sinusoidal, it is known as a harmonic load, and all of its energy is concentrated at a single frequency. A harmonic load can be expressed by a function of the form:

$$F(t) = F_0 \sin \omega t \qquad (1.24)$$

where F_0 is the load amplitude and ω its frequency. Some of the properties of harmonic motion have already been introduced in Section 1.3.2 and many of the observations made there are also applicable to harmonic loading.

Some other forms of periodic loading are shown in Figure 1.18. The third of these is a simple approximation to the load produced by a human jumping on a structure. The jumper is assumed to be in contact with the ground for 50% of the time and to impose a force in the form of a half-sine wave while in contact, and zero force while in the air.

These are nonsmooth functions, involving discontinuities in value and/or slope, which can be difficult to handle mathematically. A common way around this is to approximate them using the first few terms of a Fourier series. This is a sequence of sine and/or cosine terms, the first having the frequency of the original waveform (say $\sin \omega_0 t$) and the subsequent ones at integer multiples of this:

$$F(t) = \frac{A_0}{2} + A_1 \cos \omega_0 t + A_2 \cos 2\omega_0 t \ldots + B_1 \sin \omega_0 t + B_2 \sin 2\omega_0 t \ldots \qquad (1.25)$$

Because they are sinusoidal, the terms are again known as harmonics, with $\sin \omega_0 t$ being the first harmonic, $\sin 2\omega_0 t$ the second harmonic, and so on. The key to this, of course, is

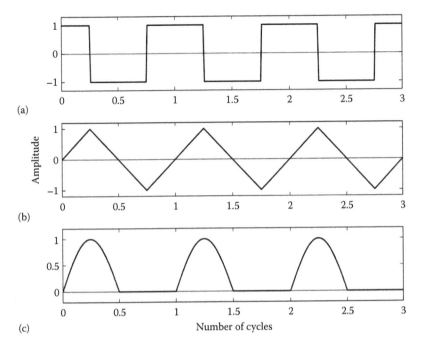

Figure 1.18 Some examples of periodic loads: (a) alternating constant amplitude, or square wave, (b) triangular wave and (c) simple representation of a load due to human jumping.

finding the right values of the harmonic coefficients A_i and B_i—how to do this is explained in Appendix C.

In most cases, Equation 1.25 can be simplified. Many functions are either *even* or *odd*, or can be made so by a simple axis shift, or by assuming a convenient extension of the function into the negative time range; in these cases, it is not necessary to include both sine and cosine terms. An even function is one that it is symmetric about the axis $t = 0$, so that $f(t) = f(-t)$. Even functions can always be represented solely by Fourier cosine terms, since $\cos t$ is itself even. Conversely, an odd function has antisymmetry about $t = 0$, so that $f(t) = -f(-t)$, and can always be represented solely by Fourier sine terms.

For example, returning to the periodic function in Figure 1.18a, if it is assumed to continue in the same form at $t < 0$ as in the range shown, then it is an even function that can be expressed as a Fourier cosine series. Similarly, the logical extension of the triangular wave in Figure 1.18b leads to an odd function, and hence a Fourier sine series. The jumping load in Figure 1.18c is neither even nor odd, but it can be made even by introducing a time shift such that the $t = 0$ axis lies either at the peak of the cosine pulse, or midway between adjacent pulses.

Figure 1.19 shows each of the functions approximated by the first four non-zero terms of their Fourier series; the coefficients used are shown in Table 1.1. It can be seen that the Fourier series converges more quickly for some periodic functions than for others—the very sharp changes in the square wave are quite difficult to represent using summed sinusoids. Nevertheless, in all cases the coefficients reduce in magnitude quite quickly as the number of the harmonic increases, so that a good enough approximation can usually be achieved using just the first few harmonics.

The key point to recognise here is that a periodic signal, which at first glance seems to be completely defined by a single frequency (or period) can in fact be represented as the

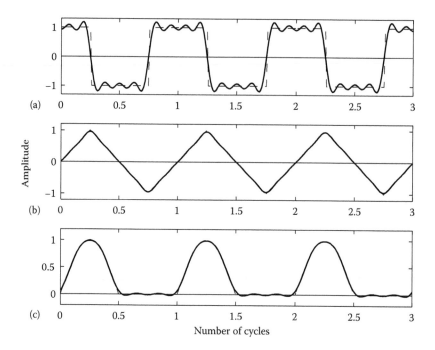

Figure 1.19 Approximation of periodic functions by Fourier series: (a) square wave, (b) triangular wave, (c) jumping load. The waveforms being approximated are shown dashed and the Fourier coefficients used are shown in Table 1.1.

Table 1.1 Fourier harmonic coefficients used in Figure 1.19

Harmonic	Constant	1	2	3	4	5	6	7
Square wave	0	1	0	−0.333	0	0.2	0	0.142
Triangular wave	0	0.811	0	−0.090	0	0.032	0	−0.017
Jumping load	0.318	0.5	0.212	0	−0.042	–	–	–

combination of a signal at that frequency and at several other higher frequencies. So, for instance, a periodic but nonharmonic load at, say, 3 Hz will necessarily have a large component at 3 Hz, but may also impart significant forces at 6 Hz, 9 Hz and 12 Hz.

1.5.3 Random loads

The load functions considered thus far are *deterministic*; we can, with quite a high degree of confidence, predict their values at some time in the future. However, many important environmental loads such as wind, ocean waves or earthquakes are *nondeterministic*, or *random*. They are irregular in form and it is impossible to predict what value they will take at some future instant in time.

An example of a random load function is shown in Figure 1.20. This shows the ground acceleration caused by the 1940 earthquake that struck southern California, as recorded in the city of El Centro. (We will see later how a record of ground acceleration can be used to determine the dynamic forces on a building.) The El Centro record is one of the earliest measured earthquake ground motions, and has become something of an industry standard in seismic analysis and testing. It can be seen that the effect of the earthquake was to accelerate

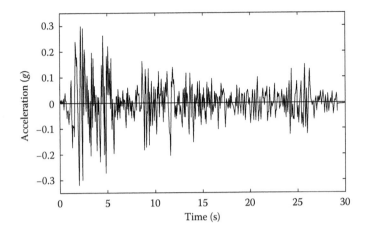

Figure 1.20 Acceleration time history recorded in the May 1940 El Centro (Imperial Valley) earthquake, north-south direction.

the ground back and forth many times over a period of about 30 seconds, the largest acceleration being around 0.32 g (3.1 m/s²). There is no regular pattern to the motion and we can be certain that, were a similar-sized earthquake to strike again at the same location, the ground acceleration record would not be the same.

Nevertheless, recording and analysis of past loading events can be extremely valuable in predicting, in general terms, the likely form of future events. If enough data are collected and analysed, we find that apparently random signals do have some common characteristics and that it is possible, by appropriate averaging and enveloping, to produce reasonably robust predictions of future events that can be used in engineering design.

Consider the El Centro earthquake record in a little more detail. It is possible to compute a frequency spectrum from the time history, shown in Figure 1.21. This has been calculated using a Fourier transform, an extension of the Fourier series approach used in Section 1.5.2 for periodic signals; Fourier transforms are discussed more fully in Chapter 6 and Appendix C. In Chapter 2, we will also introduce other ways of analysing the frequency content of an earthquake record. The spectrum shows that the acceleration time history can be broken

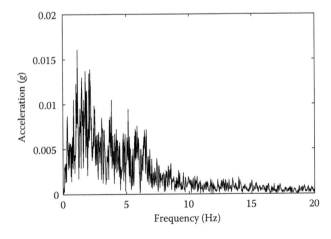

Figure 1.21 Frequency spectrum of the El Centro earthquake, calculated by Fourier transform.

down into a large number of subsignals at different frequencies, each with its own amplitude. The amplitude of a particular frequency component is unlikely to be constant over the duration of the earthquake and the spectrum shows the average value. When plotted in terms of frequency, the ground motion is still quite complex, but we can see that it is dominated by frequencies within a certain range—the strongest components are below 3 Hz, with other significant components in the range 3–7 Hz. If we repeated this analysis for other earthquakes of similar magnitude occurring in comparable seismological settings, we might expect to achieve a broadly similar spectrum, even though the time histories might look quite different.

KEY POINTS

- All structures vibrate when the loads on them change. In many instances, the motions are trivial, but some time-varying loads may result in very large vibrations. These may cause serviceability problems, structural damage or collapse.
- Many dynamic loads and motions are harmonic—varying sinusoidally with time. A good grasp of harmonic motion is a vital prerequisite to understanding behaviour that is more complex. It is particularly useful to appreciate the relationship between the sinusoidal, circular and complex exponential definitions of harmonic motion.
- Any time-varying parameter can be analysed in either the time domain, or in terms of its frequency content, by thinking of it as a superposition of harmonic components.
- The dynamic properties of a structure are dependent on its distributions of mass, stiffness and damping. Mass and stiffness provide the key ingredients for oscillatory behaviour, while damping represents energy dissipation, often in an approximate way.
- Dynamic loads can be classified as impulsive (a jolt setting the structure in motion), periodic (a regularly repeating load) or random (irregular, unpredictable loading).

TUTORIAL PROBLEMS

Where needed, take the acceleration due to gravity to be $g = 9.81$ m/s^2.

1. A harmonic vibration (in mm) is given by $x = 12 \sin \omega t + 5 \cos \omega t$. Determine the amplitude of the vibration, and the phase relative to the $12 \sin \omega t$ term.
2. An accelerometer placed on the floor of a building indicates that it is vibrating harmonically at a frequency of 20 Hz and a peak acceleration of 5 m/s^2. What is the maximum deflection of the floor during the vibration? The accelerometer is then placed on the roof and the values change to 12.5 Hz and 8 m/s^2. What is the maximum deflection of the roof?
3. Express the complex vector $z = 3 + 4i$ in the form $Ae^{i\theta}$.
4. Find the resultant of the two vibrations:

$$x_1 = 10 \sin(\omega t + 45°) \quad \text{and} \quad x_2 = 20 \sin(\omega t - 30°)$$

5. A car shock absorber is tested by applying a harmonically varying displacement of amplitude 10 mm at a frequency of 1 Hz. The shock absorber exhibits a purely viscous response with a peak resisting force of 20 N. Calculate the viscous damping coefficient. What would be the peak damper force if the frequency were increased to 5 Hz, the amplitude remaining unchanged?

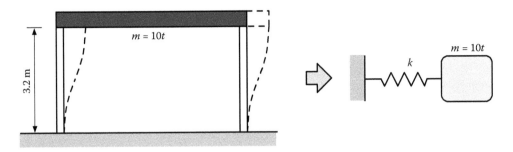

Figure P1.6

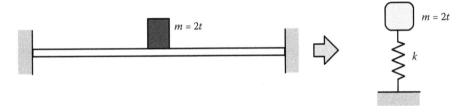

Figure P1.7

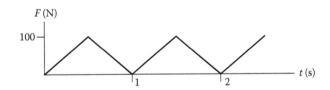

Figure P1.9

6. Figure P1.6 shows how a simple single-storey structure can be treated as an equivalent mass-spring system. The structure is idealised as a rigid roof of mass 10 tonnes supported on two flexible columns, whose mass is small compared to the roof. Each column has flexural rigidity $EI = 25 \times 10^6$ Nm^2. By assuming a deflected shape of the form shown, find the stiffness k of the equivalent mass-spring system.

7. A large, 2-tonne mass is attached to the centre of a light, flexible floor as shown in Figure P1.7. When the mass is first lowered onto the floor, it causes a static deflection of 2.8 mm. Calculate the stiffness k of the equivalent mass-spring system.

8. A man stands on a set of bathroom weighing scales on board a ship, which is moving vertically in rough seas. The reading on the scales oscillates between 50 and 90 kg with period 5.0 s. Assuming harmonic motion, estimate the mass of the man and the wave height.

9. Express the periodic force plotted in Figure P1.9 as a Fourier series, up to the term at five times the base frequency.

Chapter 2

Single-degree-of-freedom systems

2.1 INTRODUCTION

A single-degree-of-freedom (SDOF) system is one whose deformation can be completely defined by a single displacement. Obviously, most real structures have many degrees of freedom, but a surprisingly large number can be modelled approximately as SDOF systems. More importantly, it is possible to introduce most of the important concepts of dynamic analysis and behaviour by reference to SDOF systems, before moving on to more complex structures in Chapter 3.

This chapter will cover all the basic elements of the dynamic analysis and behaviour of linear SDOF systems. We will start by deriving the governing equations and explaining the key terms. We will then look at how to establish the free vibration characteristics of a structure – how it 'chooses' to vibrate in the absence of any applied dynamic loads. Consideration of the behaviour under dynamic loads will be split into two parts: first, relatively simple forms of loading (particularly harmonic), which can be described by a simple formula and where an exact, closed-form solution can be found; and second, more irregular loadcases where some form of numerical solution must be obtained. Finally, we will take a brief look at a systems-type approach, in which the dynamic behaviour is represented through block diagrams and analysed using Laplace transform methods. After reading this chapter, you should be able to

- Formulate an equation of motion for a linear SDOF system,
- Calculate its natural frequency or period,
- Appreciate the effect of damping on the free vibration behaviour,
- Calculate dynamic displacements and forces caused by step, harmonic or periodic loads,
- Calculate the response to an impulse load,
- Understand the principles of response to random loads,
- Perform a dynamic response calculation using a response spectrum,
- Understand the basics of the block diagram approach to representing dynamic systems,
- Switch between different ways of describing dynamic systems, such as structural properties, natural frequency, unit impulse response and transfer function.

2.2 EQUATION OF MOTION

2.2.1 With an applied force

Figure 2.1 shows an SDOF system comprising a mass m restrained by a spring of stiffness k and a viscous dashpot with damping coefficient c. The mass is subjected to a time-varying

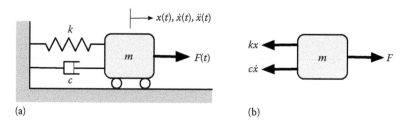

Figure 2.1 Damped SDOF system: (a) system parameters and loading and (b) free-body diagram.

external force $F(t)$, which causes a displacement $x(t)$. Using the dot notation to represent differentiation with respect to time t, the velocity is $\dot{x}$ and the acceleration is $\ddot{x}$. As the mass moves, extension of the spring causes a retarding force kx and the relative velocity across the dashpot causes a retarding force $c\dot{x}$. The forces acting on the mass at any instant are therefore as shown in the free-body diagram in Figure 2.1b. Applying Newton's second law, resultant force = mass × acceleration:

$$F(t) - kx - c\dot{x} = m\ddot{x}$$

which can be rearranged to

$$m\ddot{x} + c\dot{x} + kx = F(t) \tag{2.1}$$

Equation 2.1 is known as the *equation of motion* of the system, describing how the motion of the mass and its derivatives are related to the applied loading. For a linear system, the coefficients m, c and k are constants, so that Equation 2.1 is a linear, second-order ordinary differential equation (ODE). Much of the remainder of this chapter will involve solving this equation for different forms of the forcing function $F(t)$.

Example 2.1

A large mass M is supported at the tip of a cantilever beam of length L, and second moment of area I, made from a material with Young's modulus E. Neglecting the mass of the cantilever and any damping, derive the equation of motion for vertical vibration of the mass about its static equilibrium position.

Take downward forces and motions as positive. Split the structure at the tip of the beam, as shown in Figure 2.2. As the system vibrates, the beam and the mass exert equal and opposite forces P on each other. From simple beam theory, the force P is related to the deflection x by $P = 3EIx/L3$. Therefore, applying Newton's second law to the mass:

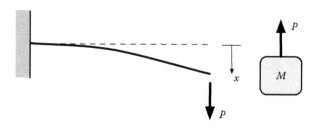

Figure 2.2 Free body diagrams for a cantilever supporting a lumped mass.

Figure 2.3 Dynamic equilibrium using d'Alembert's principle.

$$-\frac{3EI}{L^3}x = M\ddot{x} \quad \text{or} \quad M\ddot{x} + \frac{3EI}{L^3}x = 0$$

This corresponds to the standard form of equation of motion in Equation 2.1, with $c = F = 0$ and $k = 3EI/L^3$.

At first glance, it may seem odd that the gravitational force Mg has not been included in this calculation. This is because we are considering only motions *away from* the static equilibrium position. The force Mg was there before the vibration started and will still be there, unchanged, after the vibration has ceased. Thus, Mg is essential to the calculation of the static equilibrium position, but is irrelevant to vibrations about that position. The force P between the mass and the beam is the additional force over and above the static forces, which include Mg.

The way Equation 2.1 has been derived shows that motion is caused by the lack of equilibrium – the out-of-balance force results in an acceleration of the mass. An alternative way of coming to the same result is to invoke d'Alembert's principle.* Rather than considering the acceleration as a consequence of a lack of equilibrium, d'Alembert treated the $m\ddot{x}$ term as an additional internal force, called the *inertia force*, acting on the mass in the opposite direction to the acceleration (Figure 2.3). In this approach, the mass is then regarded as having an effect similar to the spring and dashpot, producing a retarding force proportional to the motion. The equation of motion is then achieved by considering equilibrium of the three internal forces with the applied load:

Inertia force + Damping force + Stiffness force = External force

$$m\ddot{x} \qquad c\dot{x} \qquad kx \qquad F(t)$$

It is, of course, important to remember that the inertia force is fictitious and the system is not really in equilibrium (otherwise, it would not accelerate). Nevertheless, d'Alembert's principle can provide an extremely useful way of thinking about the interaction of the various forces and motions within a dynamic system. To distinguish it from a normal static equilibrium, the above formulation is often referred to as an expression of *dynamic equilibrium* between the internal and external forces.

2.2.2 With a ground motion

In some forms of dynamic loading such as earthquakes, no force is applied directly to the structure; instead it is excited by a time-varying ground motion $x_g(t)$ (Figure 2.4). If we consider motions relative to a datum fixed in space, then the total motion x of the structure can be expressed as the sum of x_g and the motion of the structure relative to the ground, y:

---

* Named after the French mathematician Jean de la Rond d'Alembert (1717–1783), who first proposed it.

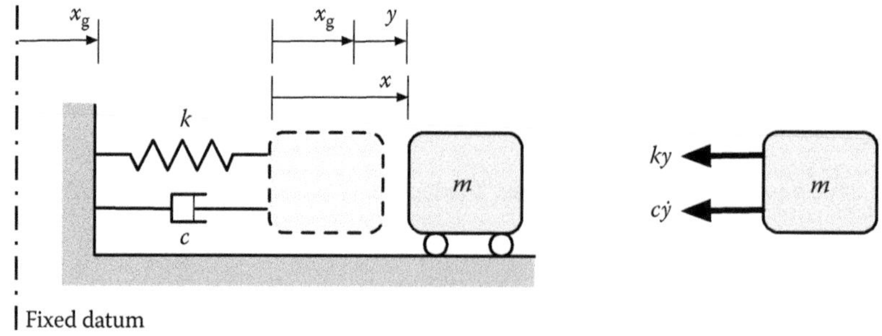

Figure 2.4 Damped SDOF system subjected to a base motion.

$$x = x_g + y \tag{2.2}$$

with a similar relationship holding for the first and second time derivatives. Now, as the mass moves from its equilibrium position, the restoring forces are proportional to the *relative* deformation across the spring and damper. However, Newton's second law tells us that the out-of-balance force gives rise to an *absolute* acceleration:

$$-c\dot{y} - ky = m\ddot{x} = m(\ddot{x}_g + \ddot{y})$$

which can be rearranged to give

$$m\ddot{y} + c\dot{y} + ky = -m\ddot{x}_g \tag{2.3}$$

So, a ground motion gives an equation of motion very similar to that due to an applied force, but in terms of relative rather than absolute motions, and with the forcing function on the right-hand side given by mass times ground acceleration.

Note that if we wish to invoke d'Alembert's principle for the ground motion case, we must remember that the inertia force is proportional to the *absolute* acceleration while the other terms are proportional to the *relative* displacement or velocity, and rearrange Equation 2.3 accordingly:

Inertia force + Damping force + Stiffness force = External force

$$m(\ddot{x}_g + \ddot{y}) \qquad c\dot{y} \qquad ky \qquad 0$$

2.3 FREE VIBRATION

Before attempting to analyse the dynamic response to an external load, it is essential to understand the free vibration of the system – how it vibrates in the absence of any external forcing. The free vibration characteristics can be considered as system properties, which can be used to simplify the solution of the forced vibration case.

2.3.1 Undamped free vibration

Consider first the theoretical case of a simple mass-spring system with no damping and no external force. The equation of motion for this case is obtained by putting both c and $F(t)$ in Equation 2.1 equal to zero, giving:

$$m\ddot{x} + kx = 0 \qquad (2.4)$$

This second order ODE can be solved by standard mathematical techniques, as set out in Appendix B. The general solution can be expressed in a variety of mathematical forms, using either the complex exponential or sine and cosine functions. One relatively simple form is

$$x = A\sin(\omega_n t + \phi) \qquad (2.5)$$

where

$$\omega_n = \sqrt{\frac{k}{m}} \qquad (2.6)$$

and the amplitude A and phase angle ϕ depend on the initial conditions. The velocity can be easily obtained by differentiating Equation 2.5 to give

$$\dot{x} = A\omega_n \cos(\omega_n t + \phi) \qquad (2.7)$$

The most common initial conditions are the imposition of an initial displacement and/or velocity. For instance, if the mass is initially held stationary at a distance X from its equilibrium position and then released, the initial conditions at $t = 0$ are $x(0) = X$ and $\dot{x}(0) = 0$. Substituting these into Equations 2.5 and 2.7 gives $A = X$, $\phi = \pi/2$ and the motion is described by

$$x = X\sin\left(\omega_n t + \frac{\pi}{2}\right) = X\cos\omega_n t \qquad (2.8)$$

Alternatively, if the mass is set in motion from its equilibrium position with a velocity V, then the initial conditions are $x(0) = 0$ and $\dot{x}(0) = V$, resulting in $A = V/\omega_n$ and $\phi = 0$, so that

$$x = \frac{V}{\omega_n}\sin\omega_n t \qquad (2.9)$$

Of course, a combination of an initial displacement and an initial velocity gives a displacement represented by the superposition of Equations 2.8 and 2.9.

The undamped free vibration behaviour therefore follows the basic rules of harmonic motion, as set out in Section 1.3.2. The mass oscillates back and forth at a constant rate and amplitude, with no loss of energy. The parameter ω_n is known as the *circular natural frequency* of the motion, normally measured in radians per second, and can be thought of as the angular velocity of an equivalent circular motion, such that one cycle of circular motion takes the same time as one vibration cycle, as illustrated in Figure 1.5.

The circular natural frequency is mathematically convenient, but more intuitive measures of the vibration rate are the *natural frequency* f_n – the number of cycles of vibration per second, and the *natural period* T_n – the time required for one complete cycle of vibration. From the circular motion analogy, one complete circle (or 2π radians) is completed in one natural period, so $\omega_n T_n = 2\pi$, or

$$T_n = \frac{2\pi}{\omega_n} = 2\pi\sqrt{\frac{m}{k}}$$

(2.10)

and

$$f_n = \frac{1}{T_n} = \frac{\omega_n}{2\pi} = \frac{1}{2\pi}\sqrt{\frac{k}{m}}$$

(2.11)

Undamped free vibrations can also be conveniently analysed using an energy approach. With no damping, energy is conserved and vibrations involve a continuous exchange between kinetic energy of the mass and strain energy in the spring. At any instant, the total energy U is

$$U = \frac{1}{2}kx^2 + \frac{1}{2}m\dot{x}^2$$

If energy is conserved, then the rate of change of U with time must be zero. Using the chain rule to differentiate the displacement and velocity terms gives

$$\frac{\mathrm{d}U}{\mathrm{d}t} = kx\dot{x} + m\dot{x}\ddot{x} = 0 \quad \text{or} \quad \dot{x}(m\ddot{x} + kx) = 0$$

Since the velocity cannot always be zero (we are only interested in systems that move!), the term in brackets must be zero and Equation 2.4 follows.

Alternatively, we can note that when the spring is at its maximum extension the velocity of the mass is zero, so that all the energy is stored as strain energy, whereas at the moment of maximum velocity the spring is unstrained, and all the energy is kinetic. Since energy is conserved

$$\frac{1}{2}kx_{\max}^2 = \frac{1}{2}m\dot{x}_{\max}^2$$

and if the motion is harmonic with displacement amplitude X then, from Equation 1.8 the velocity amplitude is $\omega_n X$, so

$$\frac{1}{2}kX^2 = \frac{1}{2}m(\omega_n X)^2 \quad \rightarrow \quad \omega_n^2 = \frac{k}{m}$$

Lastly, by dividing through by m, we can combine Equations 2.4 and 2.6 to express the equation of motion in an alternative form, where each term has units of acceleration:

$$\ddot{x} + \omega_n^2 x = 0 \tag{2.12}$$

If any equation of motion can be expressed in this general form, then the circular natural frequency of the system is simply found as the square root of the displacement coefficient.

Example 2.2

A linear spring is attached to a rigid support at its upper end. When a mass of 2 kg is attached to the bottom, the spring extends by 150 mm. If the mass is then given a small disturbance from this equilibrium position, calculate the period of the resulting oscillations.

The static force on the spring is just the weight of the mass, mg, and the stiffness is the force per unit of extension: $k = mg/x = 2 \times 9.81/0.15 = 130.8$ N/m. Then Equation 2.10 gives the period

$$T_n = 2\pi\sqrt{\frac{2}{130.8}} = 0.78 \text{ s}$$

Example 2.3

Find the natural frequency of the mass-cantilever system in Example 2.1, if the properties are $M = 100$ kg, $E = 200$ GPa, $I = 10^6$ mm^4, $L = 2.0$ m.

Dividing through by M, the equation of motion found in Example 2.1 is

$$\ddot{x} + \frac{3EI}{ML^3}x = 0$$

Comparing to Equation 2.12 gives

$$\omega_n^2 = \frac{3EI}{ML^3}$$

and substituting in the given values gives $\omega_n = 27.4$ rad/s or, dividing by 2π, $f_n = 4.4$ Hz.

2.3.2 Damped free vibration

Now consider the effect of linear viscous damping on the free vibration of our system. We return to the damped SDOF system shown in Figure 2.1, though still with the applied force set to zero, so that the equation of motion is

$$m\ddot{x} + c\dot{x} + kx = 0 \tag{2.13}$$

As with the undamped case, it is often more convenient to divide through by m to express this as an equation with units of acceleration. Recall that in Equation 1.23 we introduced a non-dimensionalised measure of damping called the damping ratio:

$$\xi = \frac{c}{2\sqrt{km}}$$

It is quite easy to see that

$$\frac{c}{m} = 2.\frac{c}{2\sqrt{km}}.\sqrt{\frac{k}{m}} = 2\xi\omega_n$$

so that Equation 2.13 can be written

$$\ddot{x} + 2\xi\omega_n\dot{x} + \omega_n^2 x = 0 \tag{2.14}$$

The way to solve this differential equation is explained in Appendix B; see in particular Example B2. Unlike the undamped case, the solution can now take one of three forms, depending on the magnitude of the damping ratio.

2.3.2.1 Critically damped: $\xi = 1$ (or $c^2 = 4\ km$)

If $\xi = 1$, then the displacement varies with time according to

$$x = (A + Bt)e^{-\xi\omega_n t} \tag{2.15}$$

where A and B are constants determined from the initial conditions. The mass returns to rest at its equilibrium position in a single half-cycle, with no oscillation (Figure 2.5a). The system is described as critically damped.

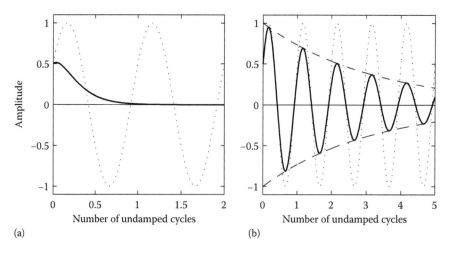

(a)   (b)

Figure 2.5 Damped free vibrations: (a) critically damped, (b) with 5% critical damping. Both cases have the same initial displacement and velocity. The undamped response is shown dashed.

The reason for the form of the damping ratio should now be clear; it takes a value of unity for the critically damped case and, for lower values of damping, it conveniently expresses the damping coefficient as a fraction of the critical value.

While it is possible to provide critical damping to a dynamic system, such a high level of damping is unusual. In mechanical systems such as cars, where viscous devices are included intentionally to damp unwanted vibrations, damping ratios of the order of 30%–40% of critical are common. In civil engineering structures, which normally rely just on the inherent damping properties of their constituent materials, much lower values are the norm – 5% is a widely assumed value for buildings, while structures such as steel bridges may possess less than 1% of critical damping.

2.3.2.2 Underdamped: $\xi < 1$ (or $c^2 < 4\ km$)

If the damping is less than the critical value, then the system is described as underdamped, and the solution to Equation 2.14 involves oscillatory motion. This is the most common case, and the one of interest to us. Like the undamped case, the motion can be expressed in several different ways, such as

$$x = Ae^{-\xi\omega_n t} \sin(\omega_d t + \phi) \tag{2.16}$$

where

$$\omega_d = \omega_n \sqrt{1 - \xi^2} \tag{2.17}$$

and the magnitude A and phase angle ϕ depend on the initial conditions. Differentiating Equation 2.16 using the product rule gives the velocity as:

$$\dot{x} = Ae^{-\xi\omega_n t}(\omega_d \cos(\omega_d t + \phi) - \xi\omega_n \sin(\omega_d t + \phi)) \tag{2.18}$$

Deriving the solutions for specified initial conditions now becomes a little more complex. With an initial displacement X and zero initial velocity, we get

$$x = \frac{X}{\sqrt{1 - \xi^2}} e^{-\xi\omega_n t} \sin(\omega_d t + \phi) \quad \text{where} \quad \tan\phi = \frac{\sqrt{1 - \xi^2}}{\xi} \tag{2.19}$$

while an initial velocity V with no initial displacement results in

$$x = \frac{V}{\omega_d} e^{-\xi\omega_n t} \sin\omega_d t \tag{2.20}$$

It is straightforward to confirm that these solutions become identical to the undamped ones (Equations 2.8 and 2.9) when $\xi = 0$.

The underdamped response differs from the undamped response in two key ways: first, energy is taken out of the system, causing an exponential decay in amplitude (Figure 2.5b); second, the frequency of the vibrations is reduced slightly, according to Equation 2.17. However, this second effect is generally of no great interest because, for practical values of damping, ω_d and ω_n take very similar values; even if the damping were 20% of critical (a very high value), the resulting damped vibration frequency would still be 98% of the undamped value.

2.3.2.3 Overdamped: $\xi > 1$ (or $c^2 > 4\ km$)

If the damping is greater than the critical value, then the system is described as overdamped. No oscillations occur and the behaviour is broadly similar to the critically damped case. This case need not concern us further.

2.3.2.4 Logarithmic decrement

Another measure of damping that is often used is the logarithmic decrement. For a viscously damped system, the exponential decay of vibrations means that the ratio between any two successive peaks is the same. Referring to Figure 2.6, the logarithmic decrement, δ, is simply the natural log of this ratio:

$$\delta = \ln\frac{X_1}{X_2} = \ln\frac{X_2}{X_3} = \ln\frac{X_n}{X_{n+1}}$$
(2.21)

where n is any positive integer. At the peaks, the sine function takes a value of unity so Equation 2.16 gives

$$X_1 = Ae^{-\xi\omega_n t_1}, \quad X_2 = Ae^{-\xi\omega_n t_2} = Ae^{-\xi\omega_n (t_1+T_d)}$$

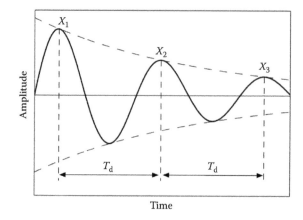

Figure 2.6 Viscously damped free vibrations giving a constant amplitude ratio between successive peaks.

Therefore, substituting into Equation 2.21 and remembering that $\omega_d \approx \omega_n$ for all practical values of ξ, we get

$$\delta = \xi\omega_n T_d = \xi\omega_n . \frac{2\pi}{\omega_d} \approx 2\pi\xi \tag{2.22}$$

Clearly, as with period and frequency, the various definitions of damping are interchangeable and there is no particular reason to prefer one to another. Nevertheless, engineers working in different disciplines have adopted different preferences. For instance, earthquake engineers generally define oscillation rate using the period and damping using the damping ratio, whereas wind engineers often prefer to use frequency and logarithmic decrement.

Example 2.4

In a vibration test on a building floor, the floor is set in motion by a blow from an impact hammer and its free vibrations monitored using an accelerometer. It is found to have a natural frequency of 6.0 Hz, and its vibration amplitude reduces to 10% of its initial value in 1 s. Calculate the damping ratio.

One second corresponds to 6 cycles of vibration, during which time the amplitude reduces by a factor of 10. Assuming purely viscous damping, the amplitude reduces by the same ratio in each cycle and so

Ratio of cycle n to cycle $(n + 6)$: $\dfrac{X_n}{X_{n+6}} = \left(\dfrac{X_n}{X_{n+1}}\right)^6 = 10$

The log decrement $\delta = \ln\dfrac{X_n}{X_{n+1}} = \ln(10^{1/6}) = \dfrac{1}{6}\ln(10)$

and the damping ratio is $\xi = \dfrac{\delta}{2\pi} = \dfrac{\ln(10)}{12\pi} = 0.061$, or 6.1% of critical.

2.4 RESPONSE TO DYNAMIC LOADS

Once the free vibration properties of a system have been determined, it is possible to go on and analyse how the structure will respond to the application of a time-varying external force. For an SDOF system, this is given by the solution of the equation of motion, Equation 2.1:

$$m\ddot{x} + c\dot{x} + kx = F(t)$$

When looking at free vibrations, we noted that it is often helpful to divide through by the mass and replace c and k by ξ and ω_n. Applying this to Equation 2.1 gives

$$\ddot{x} + 2\xi\omega_n\dot{x} + \omega_n^2 x = \frac{F(t)}{m} \tag{2.23}$$

or, for the base motion case, Equation 2.3 becomes

$$\ddot{y} + 2\xi\omega_n\dot{y} + \omega_n^2 y = -\ddot{x}_g \tag{2.24}$$

The general solution method for differential equations of this sort is set out in Appendix B and will not be presented in depth in the main text. It is important to note, however, that the full solution is always made up of two parts – the first obtained by setting the right-hand side to zero (i.e. solving the free vibration problem) and the second by substituting in a trial solution whose form is based on that of $F(t)$. In mathematical terminology:

Full solution = Complementary function + Particular integral,

or, in dynamics terminology:

Full solution = Transient response + Steady state response.

The *transient* response is a vibration at the natural frequency of the structure. It can be thought of as a free vibration initiated by the onset of the applied load, which disturbs the structure from its equilibrium position. It is called transient because the damping causes it to die away quite quickly. In relatively short-duration events, the transient response can be a significant part of the total, but it is often neglected when considering long-duration loads.

The nature of the *steady state* response will vary with that of the applied loading, and will continue for as long as the loading.

In the following sections, we will consider the dynamic behaviour of SDOF systems subjected to a variety of applied loads.

2.4.1 Step load

To introduce the idea of dynamic response to a load, we start with a very simple case. A mass-spring-damper system is initially at rest at its equilibrium position. At time zero, the load applied to the mass is suddenly increased from zero to some value F_0, and then held constant, as shown in Figure 2.7a. The transient response is just the free vibration solution, given by Equation 2.16. The steady state response to a constant load is easily found, and unsurprisingly is a constant displacement: $x = F_0/k$. The full solution is then

$$x = Ae^{-\xi\omega_n t}\sin(\omega_d t + \phi) + \frac{F_0}{k} \tag{2.25}$$

and substituting in the initial conditions, $x(0) = \dot{x}(0) = 0$, results in the following expressions for A and ϕ:

$$A = -\frac{F_0}{k\sqrt{1-\xi^2}}, \quad \tan\phi = \frac{\sqrt{1-\xi^2}}{\xi} \tag{2.26}$$

The response is plotted in Figure 2.7b–e for three different values of damping, as well as for the undamped case. Also shown dashed is the displacement $x_0 = F_0/k$ that would result if F_0 were applied as a static load. In this case, the static response is the same as the steady state part of the dynamic response, and the oscillations about this are due to the transient response. For damped systems, these oscillations die out, leaving just the long-term displacement x_0.

This sort of dynamic response can be seen when you stand on a set of bathroom scales. The scales determine your weight through the deformation of a stiff spring. Initially the

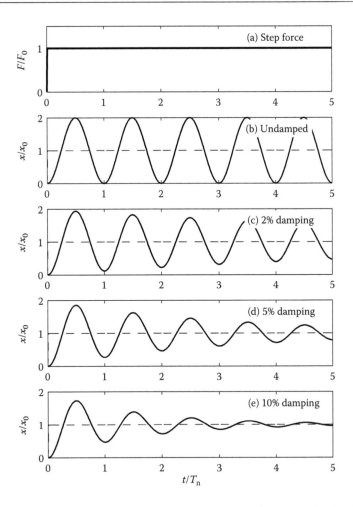

Figure 2.7 Vibration of an SDOF system in response to a step load, for various levels of damping (static response x_0 shown dashed).

displacement, and hence the weight indicated, fluctuates due to the transient response of the system. However, the vibrations are quickly damped out, leaving the scales displaying just your static weight.

From Figure 2.7 it can be seen that the peak dynamic displacement is significantly greater than the static one, by a factor of 2 for the undamped case and by smaller factors when damping is present. This factor is known as the dynamic amplification factor (DAF), D, and provides an extremely useful measure of the severity of a dynamic response.

$$D = \frac{x_{max}}{x_0} \tag{2.27}$$

For the step load response, x_{max} always occurs at the first positive peak, is independent of the natural frequency of the system and depends only on the damping, as shown in Figure 2.8. Thus, a lightly damped structure subjected to a suddenly applied load

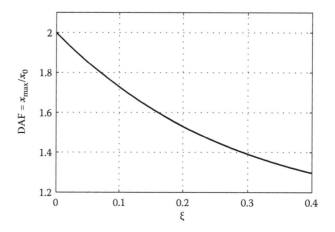

Figure 2.8 Dynamic amplification factor for an SDOF system responding to a step load, as a function of the damping ratio.

can briefly deflect up to twice as far as would be predicted by a simple static deflection calculation.

2.4.2 Harmonic load

We can now move on to a much more interesting, and onerous, load case. A harmonic load is one that varies sinsoidally with time and continues at constant amplitude for many cycles. Many items of machinery contain large rotating masses, which impose centripetal forces on the supporting structure as they rotate. Recalling the equivalence between harmonic and circular motion, it is easy to see that resolving these rotating forces into horizontal and vertical components results in harmonic forces. If we define the harmonic force as

$$F(t) = F_0 \sin \omega_F t$$

where ω_F is the circular forcing frequency, then the equation of motion can be written:

$$\ddot{x} + 2\xi\omega_n\dot{x} + \omega_n^2 x = \frac{F_0}{m}\sin \omega_F t \qquad (2.28)$$

Since it is obtained by putting the right-hand side (RHS) of Equation 2.28 equal to zero, the transient part of the solution will once again take the general form given in Equation 2.16. The more significant part is the steady state solution. As explained in Appendix B, we obtain this by substituting a function having the same general form as the loading function.

$$x = x_{max} \sin(\omega_F t - \phi)$$

More conveniently, we can express this as

$$x = Dx_0 \sin(\omega_F t - \phi) \qquad (2.29)$$

where D is the dynamic amplification factor and $x_0 = F_0/k$ is the static displacement that would be caused by a constant force F_0. After substituting into Equation 2.28 and a fair amount of mathematical manipulation, the expressions for D and ϕ are found to be

$$D = \frac{x_{max}}{x_0} = \frac{1}{\sqrt{[1-\Omega^2]^2 + [2\xi\Omega]^2}}$$

(2.30)

$$\tan\phi = \frac{2\xi\Omega}{1-\Omega^2}$$

(2.31)

where

$$\Omega = \frac{\omega_F}{\omega_n} = \frac{T_n}{T_F}$$

(2.32)

The frequency ratio Ω takes a low value when the load is applied slowly and/or the structure is very stiff (high natural frequency), is unity when the loading and natural frequencies match, and takes a high value when the loading oscillates rapidly and/or the structure is very flexible (low natural frequency).

Figure 2.9 shows how the DAF and phase angle vary with the frequency ratio. The behaviour can be divided into three distinct regimes:

- At the left-hand end ($\Omega \ll 1$) the loading is slow compared to the natural oscillations of the structure. Here, the response is *quasi-static* – the DAF is close to unity and the phase angle is approximately zero.
- As we approach a frequency ratio of 1, we are pushing the mass to and fro at about the rate it naturally tends to oscillate anyway. As a result, it offers little resistance and we get very large amplification of the motion compared to the quasi-static case – this is *resonance*. At $\Omega = 1$, the phase angle is always 90° ($\pi/2$ rads) and the DAF is

$$D_{\Omega=1} = \frac{1}{2\xi}$$

(2.33)

- As we move to the right-hand end of the graphs ($\Omega \gg 1$), the motion of the mass gets smaller and smaller – even less than under the static load F_0.

To understand these different forms of response a little better, it can be helpful to look at how the motions vary over time. Figure 2.10 shows time histories of the motion for each of the three regimes. Each plot shows the displacement of a 5%-damped SDOF system subjected to a harmonic load of unit magnitude. The system starts from rest and the solution shown includes the initial transient part (not presented in detail above) as well as the steady state response; the transient becomes negligible by about halfway through the plots, and the system settles down to just the steady state. For comparison, the thin line shows the displacement $F(t)/k$ that would result if the dynamic terms due to inertia and damping were ignored.

Figure 2.10a corresponds to $\Omega = 0.25$, meaning that the load is applied quite slowly. Here it can be seen that the dynamic analysis adds very little to the static one. The transient

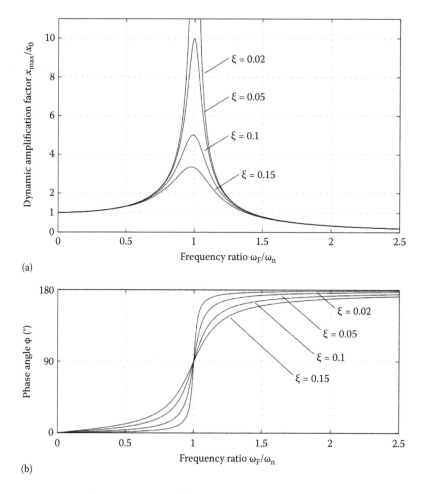

Figure 2.9 (a) Dynamic amplification factor and (b) phase angle for steady state response of an SDOF system to a harmonic force, as a function of the frequency ratio.

part of the response adds a little ripple to the first cycle or so, but once that damps out, the static and dynamic responses are virtually identical, with no significant time lag between them.

Figure 2.10b shows the resonant case, $\Omega = 1$. Note that the vertical scale differs from the other plots by a factor of 10, to accommodate the much greater motion. Here, as the transient decays away, the amplitude builds up until it is, in this case, 10 times the static one (as can be seen by putting $\xi = 0.05$ in Equation 2.33). The response lags the applied loading by 90° – recalling that the velocity of a harmonic motion leads the displacement by 90° (see Section 1.3.2), this means that the velocity is exactly in phase with the applied force. Therefore, the peak positive force is being applied to the mass at the moment when its displacement is zero and its velocity is maximum. Thus, the force is always pushing the mass in the direction it is already moving, making it easy to amplify the motion. This is rather like pushing a child on a swing; to make them swing higher, you must push at the point in the swing cycle when their velocity is away from you.

In Figure 2.10c we are well beyond the resonant peak, at $\Omega = 2$. Here, once the transient has been damped out, we are left with a rather small oscillation, in anti-phase with the

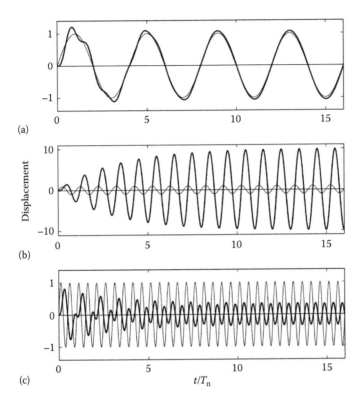

Figure 2.10 Time histories of displacement response of a 5% damped SDOF system to a unit-amplitude harmonic load. The frequency ratio is (a) 0.25 – quasi-static loading, (b) 1.0 – resonance, and (c) 2.0 – fast loading. The thin line is the static displacement (*F/k*) and the heavy line the dynamic response.

loading, that is, the peak positive force is applied at the instant of peak negative displacement, and vice versa.

Further insight can be gained from Figure 2.11, which shows the relative contributions of the inertia force, the damping force and the stiffness force, in both time domain and phasor formats. We saw in Chapter 1 that these three force components are proportional to the acceleration, velocity and displacement, respectively, and are always separated by phase angles of 90° each; this means that the displacement and acceleration are always in antiphase, and that when they reach their extreme values the velocity is always zero.

For the quasi-static case (Figure 2.11a), the velocity and acceleration are both low, and therefore so are the damping and inertia forces. As a result, almost all of the applied force is balanced by the stiffness term, and the phase angle between the applied force and the stiffness force is very small.

At resonance (Figure 2.11b), all three force components are significant, but the inertia and stiffness terms are always exactly equal and opposite, cancelling each other out. The only resistance to the applied load comes from the damping force; this explains why the height of the resonant peak in Figure 2.9a depends only on the damping level. Note also that the damping force is now in phase with the applied load.

Lastly, when $\Omega > 1$ the inertia force becomes the largest, and becomes the one closest to being in phase with the applied load (Figure 2.11c). As a result, the phase angle between the load and the stiffness force (and hence the displacement) approaches 180°.

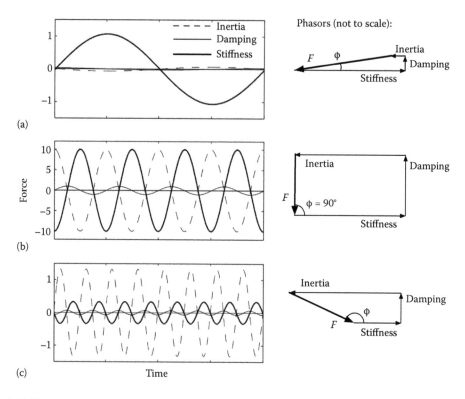

Figure 2.11 Time histories and phasor representation (not to scale) of inertia, damping and stiffness force components for a 5% damped SDOF system subjected to a unit-amplitude harmonic load, with frequency ratio (a) 0.25, (b) 1.0, (c) 2.0.

Example 2.5

A large rotary machine has mass 2 tonnes. When placed on a suspended floor, it produces a static deflection of 1.8 mm. At its operating speed of 1000 rpm, an out-of-balance force of amplitude 5 kN is generated. Assuming a damping ratio of 3%, calculate the resulting steady state vibration amplitude of the floor.

The static floor deflection is mg/k, therefore $k = 2000 \times 9.81/0.0018 = 10.9 \times 10^6$ N/m. Assuming that the mass of the floor is small compared to that of the machine, the natural frequency is

$$\omega_n = \sqrt{\frac{k}{m}} = \sqrt{\frac{10.9 \times 10^6}{2000}} = 73.8 \text{ rad/s}, \quad \text{or} \quad f_n = 11.8 \text{ Hz}$$

The rotating unbalanced mass will cause a centripetal force whose vertical component varies sinusoidally with time. A speed of 1200 rpm can be expressed as

$\omega_F = 1200 \times 2\pi/60 = 125.7$ rad/s, giving a frequency ratio $\Omega = 125.7/73.8 = 1.7$.

Together with $\xi = 0.03$, this results in a DAF from Equation 2.30 of $D = 0.53$.

The vibration amplitude is then $DF_0/k = 0.53 \times 5000/10.9 \times 10^6 = 0.24$ mm.

Although this looks like quite a small deflection, it corresponds to an acceleration amplitude of $\omega^2 x = 3.8$ m/s², which would most likely be unacceptable. The machine would need to be mounted on a vibration isolation support.

Note also that a worse response would be achieved with the machine running at a lower speed of 73.8 rad/s, inducing resonance. This would give $D = 16.7$ and a vibration amplitude of 7.6 mm. The machine must briefly pass through this frequency as it accelerates up to its operating speed at start-up. However, this should be acceptable so long as the frequency is not maintained for too long, as the resonant response takes some time to build up (as you can see in Figure 2.10b). You can notice a similar sort of behaviour in an ordinary domestic washing machine. As it enters the spin cycle, there is briefly a resonance between the rotation of the out-of-balance mass of the wet clothes and the natural frequency of the spring-mounted drum containing them; often, a knocking sound can be heard. However, as the rotational speed of the drum increases rapidly away from the resonant speed, a stable spin is achieved and the knocking stops.

2.4.3 Harmonic base motion

The general equation of motion for a structure subjected to a base motion was given in Equation 2.24. If the base motion takes the form $x_g = X_g \sin \omega_g t$, then we have

$$\ddot{y} + 2\xi\omega_n\dot{y} + \omega_n^2 y = -\ddot{x}_g = \omega_g^2 X_g \sin \omega_g t \tag{2.34}$$

where $y = x - x_g$ is the displacement of the structure relative to the ground. As might be expected, the solution for this case has many similarities to the harmonic force case. The solution can be given either for the relative motion y or the absolute motion x. Here we present the latter. The steady state response can be written as

$$x = DX_g \sin(\omega_g t - \phi) \tag{2.35}$$

where the dynamic amplification factor is now defined as the ratio of the structure displacement amplitude to that of the ground. The expressions for D and ϕ in this case are

$$D = \frac{x_{max}}{X_g} = \sqrt{\frac{1+[2\xi\Omega]^2}{[1-\Omega^2]^2 + [2\xi\Omega]^2}} \tag{2.36}$$

$$\tan\phi = \frac{2\xi\Omega^3}{[1-\Omega^2] + [2\xi\Omega]^2} \tag{2.37}$$

where

$$\Omega = \frac{\omega_g}{\omega_n} = \frac{T_n}{T_g} \tag{2.38}$$

Figure 2.12 shows how the DAF and phase angle vary with the frequency ratio. The behaviour can again be divided into three distinct regimes. At slow rates of loading ($\Omega \ll 1$) the structure just moves with the ground, without any deformation, that is, $x = x_g$. At a frequency ratio close to 1 we again get a resonant response, though there are some slight differences to the applied force case; at $\Omega = 1$ the phase angle is somewhat less than 90° ($\pi/2$ rads) and the DAF is

$$D_{\Omega=1} = \frac{\sqrt{1+[2\xi]^2}}{2\xi} \tag{2.39}$$

This is quite close to $1/(2\xi)$ for usual levels of damping. As we move to the right-hand end of the graphs ($\Omega \gg 1$), the motion of the mass gets very small – to a first approximation, the mass stays still as the ground moves beneath it.

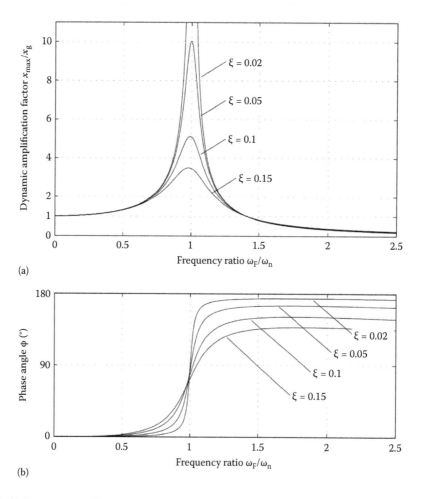

(a)

(b)

Figure 2.12 (a) Dynamic amplification factor and (b) phase angle for steady state response of an SDOF system to a harmonic base motion, as a function of the frequency ratio.

Example 2.6

A car is (very crudely) idealised as an SDOF system with mass $m = 1.2$ tonnes, suspension stiffness $k = 180$ kN/m. The shock absorbers provide 30% of critical damping. The car is driven along a road with an undulating profile, idealised as sinusoidal with amplitude 10 mm and wavelength 10 m. Describe the vertical motion felt by the driver at speeds of 10, 20 and 30 m/s.

The vertical natural frequency of the vehicle is $f_n = \dfrac{1}{2\pi} \sqrt{\dfrac{180 \times 10^3}{1200}} = 1.95\,\text{Hz}$

The analysis then proceeds through the following steps for each velocity – the results are tabulated below.

1. If the surface profile has wavelength λ_g and the vehicle velocity is v, then the frequency at which peaks are passed is $f_g = v/\lambda_g$.
2. The frequency ratio $\Omega = f_g/f_n$.
3. The DAF can then be calculated by substituting Ω and $\xi = 0.3$ into Equation 2.36.
4. The displacement amplitude is then DX_g.
5. The acceleration amplitude is obtained by multiplying this value by $\omega_g^2 = (2\pi f_g)^2$.

	Velocity	10	20	30	(m/s)
1.	$f_g = v/\lambda_g$	1.0	2.0	3.0	(Hz)
2.	$\Omega = f_g/f_n$	0.51	1.03	1.54	
3.	D – Equation 2.36	1.31	1.89	0.82	
4.	$x_{max} = DX_g$	6.5	9.5	4.1	(mm)
5.	$\ddot{x}_{max} = x_{max}(2\pi f_g)^2$	0.26	1.5	1.5	(m/s^2)

Note that the DAFs are comparatively small, even at resonance, because of the high damping – typical of a car but much higher than would normally be expected in a structure. While the resonant (20 m/s) case gives the higher displacement, the acceleration is approximately equal at 20 and 30 m/s. In general, higher frequency vibrations will involve higher accelerations. For ride comfort, the acceleration is the more important parameter. The peak values achieved here are likely to be acceptable so long as they are only sustained for a short time.

Example 2.7

A single-storey frame is idealised as an SDOF system with mass 10 tonnes, natural period 0.1 s and damping ratio 5%. Ground vibrations caused by a nearby excavation site are approximated as sinusoidal with frequency $f_g = 10$ Hz and amplitude $X_g = 0.2$ mm. Estimate the dynamic response of the structure at roof level.

The structural properties given imply stiffness and damping coefficients for the structure of

$$k = (2\pi)^2 \frac{m}{T_n^2} = 39.5\,\text{MN/m}, \quad c = \xi.2\sqrt{km} = 62.8\,\text{kNs/m}$$

The ground displacement (in m) can be written as

$$x_g = 0.0002 \sin(20\pi t)$$

Although the displacement amplitude is small, the frequency of the vibration is relatively high, resulting in significant velocities and accelerations. The velocity amplitude is $20\pi \times 0.0002 = 0.0126$ m/s and the acceleration amplitude is $(20\pi)^2 \times 0.0002 = 0.79$ m/s^2.

The natural period of the structure corresponds to a frequency of 10 Hz, so the frequency ratio is 1. Equation 2.36 then gives $D = 10.05$ and Equation 2.37 gives $\phi = 1.47$ rads = 84.3°. Thus, the absolute displacement x of the roof can be written as

$$x = DX_g \sin(\omega_g t - \phi) = 0.00201 \sin(20\pi t - 1.47)$$

and differentiating gives

$$\dot{x} = 0.00201 \times 20\pi \cos(20\pi t - 1.47)$$

$$\ddot{x} = -0.00201 \times (20\pi)^2 \sin(20\pi t - 1.47)$$

x has an amplitude just over 10 times that of the ground, and is just less than 90° out of phase with the ground, so that the peak roof displacement occurs just before the zero ground displacement. As shown in Figure 2.13, this results in a displacement y relative to the ground whose amplitude is exactly $1/(2\xi) = 10$ times that of the ground and which is exactly 90° out of phase with the ground.

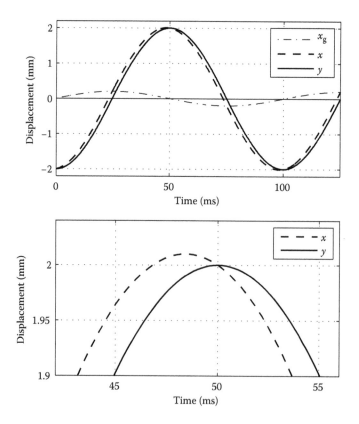

Figure 2.13 A typical part of the absolute (x) and relative (y) steady state displacements of an SDOF structure in response to a harmonic base motion x_g. The lower plot is a zoom of the peak, to show more clearly the small amplitude and phase differences between x and y.

Now consider the internal forces at various instants during the response, remembering that the relative displacement and velocity give rise to the stiffness and damping forces, while the absolute acceleration corresponds to the inertia force.

$t = 25$ ms: x_g takes its maximum value of 0.2 mm. As can be seen from Figure 2.13, x also has a value of 0.2 mm, so that $y = 0$. It follows that the stiffness force ky is zero. Since the ground velocity is zero at this instant, we have $\dot{y} = \dot{x}$. Substituting in numerical values then results in equal and opposite damping and inertia forces:

$$c\dot{y} = 62800 \times 0.126 = 79.0 \text{ kN}$$

$$m\ddot{x} = 10000 \times -0.79 = -79.0 \text{ kN}$$

$t = 50$ ms: Again referring to Figure 2.13, $x_g = 0$, and $x = y = 2.0$ mm. At this instant, the relative velocity and hence the damping force are zero and we have equal and opposite stiffness and inertia forces:

$$ky = 39.5 \times 10^6 \times 0.002 = 789.6 \text{ kN}$$

$$m\ddot{x} = 10000 \times -7.896 = -789.6 \text{ kN}$$

In general, the second instant is the one of design interest – it gives the largest stiffness force, which the columns of the structure must be designed to withstand. The largest stiffness force always occurs at the instant of maximum displacement relative to the ground, at which instant the relative velocity (and hence the damping force) must be zero, so that there is a dynamic equilibrium between the stiffness and inertia forces.

Note that it is possible to find a slightly larger inertia force. At $t = 48.4$ ms the absolute displacement and acceleration are both very slightly greater than at $t = 50$ ms and an inertia force of –793.5 kN is achieved. However, at this instant, the relative displacement is slightly below its maximum value, giving a somewhat lower stiffness force, the balance being made up by a small damping force.

2.4.4 Response to periodic loads

Periodic loads repeat themselves at regular intervals. A harmonic load is a periodic load comprising a single frequency component, which can therefore be described mathematically using a single sine or cosine function. For other types of periodic load, we saw in Section 1.5.2 that a Fourier series can be used to express them as the sum of an average load and a series of sine and/or cosine terms, known as harmonics.

Analysis of the response to a periodic load is therefore a straightforward extension of the harmonic load case. The response to each harmonic term in the series can be found using the harmonic load approach, and the total response is simply obtained by summing the responses to each term.

If a periodic load repeats itself with period T_F, then its Fourier series takes the form:

$$F(t) = \frac{A_0}{2} + \sum_{n=1}^{\infty} (A_n \cos n\omega_F t + B_n \sin n\omega_F t) \qquad (2.40)$$

where $\omega_F = 2\pi/T_F$. Appendix C explains how to calculate the coefficients A_n and B_n, and gives their values for common periodic functions. The constant term is expressed as $A_0/2$ as this allows it to be determined using the same formula as the other A_n. In most cases, a periodic function can be represented by either a sine or a cosine series – it is not particularly common to require both sets of terms. Consider first the example of an odd function (i.e. one for which $f(-t) = f(t)$), defined by a Fourier sine series.

For a linear SDOF system, the displacement caused by the constant force is simply achieved by dividing by the stiffness k. For each sine term, the steady state response can be found from Equation 2.29, with the static displacement x_0 given by B_n/k:

$$x_n = D_n \frac{B_n}{k} \sin(n\omega_F t - \phi_n)$$

(2.41)

where the DAF D_n and phase angle ϕ_n can be found from Equations 2.30 and 2.31. Summing for all the terms in the series, the total response to the periodic load is then:

$$x(t) = \frac{1}{k}\left(\frac{A_0}{2} + \sum_{n=1}^{\infty} D_n B_n \sin(n\omega_F t - \phi_n) \right)$$

(2.42)

Obviously, a similar expression can be written for the response to a load function made up of Fourier cosine terms, replacing sine by cosine and B_n by A_n:

$$x(t) = \frac{1}{k}\left(\frac{A_0}{2} + \sum_{n=1}^{\infty} D_n A_n \cos(n\omega_F t - \phi_n) \right)$$

(2.43)

As when approximating a load function using a Fourier series, it will usually be sufficient to include just the first few terms of the summation.

Example 2.8

A person jumping regularly on a floor exerts a periodic force that can be approximated as shown in Figure 2.14a; when in contact with the floor the vertical force varies as a cosine function with a peak magnitude of 2.2 kN. They perform two jumps per second and are in contact with the floor for 50% of the time. The floor is modelled as an SDOF system with mass 4.8 tonnes, natural frequency 4.5 Hz and 2% of critical damping. Using a Fourier series approximation to the loading, estimate the peak deflection and acceleration of the floor.

The circular frequency of the loading is $\omega_F = 2\pi f_F = 2\pi \times 2 = 4\pi$ rad/s

The circular natural frequency of the floor is $\omega_n = 2\pi f_n = 2\pi \times 4.5 = 9\pi$ rad/s

The stiffness of the floor is $k = \omega_n^2 m = (9\pi)^2 \times 4800 = 3.838 \times 10^6$ N/m

The idealised load function comprises two half-cosine waves per second, each of duration 0.25 s and separated by a period of 0.25 s when the load is zero (the jumper is in the air). From Appendix C, Table C.1, the first four terms of the Fourier series for this load are

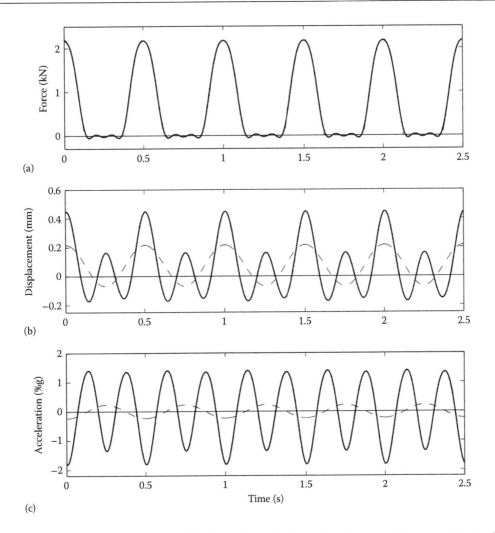

Figure 2.14 Effect of jumping on a floor: (a) idealised jumping force time history and its approximation by Fourier series terms up to the fourth harmonic, (b) resulting floor displacement, (c) floor acceleration. In (b) and (c), the solid line is calculated using terms up to the fourth harmonic, the dashed line using the first harmonic only.

$$F(t) = F_{max}\left(\frac{1}{\pi} + \frac{1}{2}\cos\omega_F t + \frac{2}{3\pi}\cos 2\omega_F t - \frac{2}{15\pi}\cos 4\omega_F t \ldots\right)$$

$$= 2200\left(\frac{1}{\pi} + \frac{1}{2}\cos 4\pi t + \frac{2}{3\pi}\cos 8\pi t - \frac{2}{15\pi}\cos 16\pi t \ldots\right)$$

This equation comprises first, second and fourth harmonics; in this instance, the coefficient of the third harmonic turns out to be zero. The constant term represents the average load, which in this case is just the static weight of the jumper ($2200/\pi = 700$ N). Figure 2.14a shows the idealised load and its Fourier approximation. An excellent match is achieved during the load regions, with a slight ripple visible in the approximation of the zero-load regions.

For each of the three harmonic terms, calculate the DAF D_n and phase angle ϕ_n using Equations 2.30 and 2.31:

First harmonic: $\quad \Omega = \dfrac{2}{4.5} = 0.444 \quad \rightarrow \quad D_1 = 1.25, \quad \phi_1 = 0.022 \text{ rads}$

Second harmonic: $\Omega = \dfrac{4}{4.5} = 0.889 \quad \rightarrow \quad D_2 = 4.70, \quad \phi_2 = 0.168 \text{ rads}$

Fourth harmonic: $\quad \Omega = \dfrac{8}{4.5} = 1.778 \quad \rightarrow \quad D_4 = 0.46, \quad \phi_4 = 3.109 \text{ rads}$

Substituting these values into Equation 2.43 gives the full displacement solution:

$$x(t) = \frac{2200}{k}\left(\frac{1}{\pi} + \frac{1.25}{2}\cos(4\pi t - 0.022) + \frac{2 \times 4.70}{3\pi}\cos(8\pi t - 0.168) \right.$$

$$\left. - \frac{2 \times 0.46}{15\pi}\cos(16\pi t - 3.109)... \right)$$

The resulting displacement is plotted in Figure 2.14b. The solid line shows the full solution achieved by the above equation. To show the importance of the higher Fourier terms, the dashed line shows the solution that would be achieved if only the first harmonic were included. In this case, the second harmonic (which is the closest to resonance) has a significant effect on the displacement response, while the fourth harmonic is very small and makes a negligible contribution. As would be expected for a single person jumping on a substantial floor, the displacement amplitude is small.

To find the accelerations, each harmonic term must be differentiated twice, equivalent to multiplying by minus the square of its frequency, giving

$$\ddot{x}(t) = -\frac{2200}{k}\left((4\pi)^2 \frac{1.25}{2}\cos(4\pi t - 0.022) + (8\pi)^2 \frac{2 \times 4.70}{3\pi}\cos(8\pi t - 0.168) \right.$$

$$\left. - (16\pi)^2 \frac{2 \times 0.46}{15\pi}\cos(16\pi t - 3.109)... \right)$$

The acceleration is plotted in Figure 2.14c, with the dashed curve again showing the result achieved by using just the first harmonic term. Because they are achieved by multiplying by the frequency squared, the higher harmonics can often give rise to quite high accelerations, and so can be worth including even if they give low displacements. In this case, even with the frequency multiplier, the pattern is broadly similar to the displacements – the second harmonic is significant but the fourth makes only a small contribution to the overall response. The acceleration achieved is of the order of 1% g (i.e. about 0.1 m/s^2), which is not dangerous to the structure but may cause annoyance to building occupants if sustained for a long time.

2.4.5 Impulse response

Although the term *impulsive* is often used to refer to various types of load involving sudden changes, a true impulse load is one having very short duration and, usually, large magnitude. Loads of this sort can arise due to high-velocity impact or explosive blast but, as we shall see, the usefulness of this load case transcends any particular application because it provides a fundamental building block for the analysis of more complex load cases.

If a load has extremely short duration, then we are less concerned with its magnitude than with the impulse it provides, that is, the integral of its magnitude over its duration. To help us define this neatly, we use a mathematical function called the Dirac delta function, $\delta(t-a)$. The delta function takes a value of zero at all values of t except $t = a$, where it takes a very large, undefined value. However, its value at $t = a$ is assumed to be such that its time integral is one. Expressed mathematically

$$\delta(t-a) \begin{cases} = 0 & t \neq a \\ \rightarrow \infty & t = a \end{cases} \text{ and } \int_{-\infty}^{\infty} \delta(t-a)\,dt = 1 \tag{2.44}$$

We will consider this function in more depth in Chapter 6.

A short-duration force occurring at time $t = a$ and resulting in an impulse of magnitude I can be written as $I\delta(t-a)$. The equation of motion of a system subjected to an initial impulse I then becomes

$$\ddot{x} + 2\xi\omega_n\dot{x} + \omega_n^2 x = \frac{I}{m}\delta(t) \tag{2.45}$$

This can be solved using the method of Laplace transforms (see Appendix D for a brief introduction) or more simply by recognising that, because the impulse is so short, it can be thought of as an initial condition rather than a continuous load function. Assume the system is initially at rest at its equilibrium position. An impulse acting on a system causes a change of momentum (mass × velocity). Therefore, an impulse I instantaneously imparts a velocity I/m to the mass and, the loading now having finished, the mass simply undergoes free vibrations subject to the initial conditions:

$$x(0) = 0, \quad \dot{x}(0) = \frac{I}{m}$$

Substituting these conditions into Equations 2.5 and 2.7 allows us to solve for the amplitude and phase of the vibration, giving

$$x = \frac{I}{m\omega_d}e^{-\xi\omega_n t}\sin\omega_d t \tag{2.46}$$

Of course, if the impulse is applied at some time other than zero, say $t = a$, then Equation 2.46 still applies, just replacing t by $(t-a)$.

The response of a system to a unit impulse ($I = 1$) is often used as an alternative way of defining its dynamic properties, and in this context is usually denoted by $h(t)$:

$$h(t) = \frac{1}{m\omega_d}e^{-\xi\omega_n t}\sin\omega_d t \tag{2.47}$$

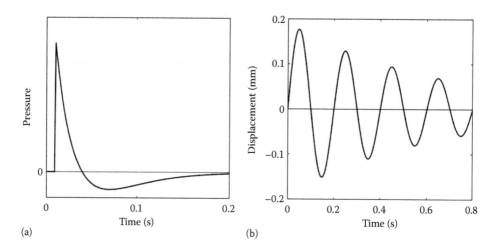

Figure 2.15 Response of an SDOF structure to a blast wave: (a) typical blast pressure time history (idealised as an instantaneous impulse in this analysis), (b) displacement response.

Example 2.9

An explosive blast results in a very sharp rise in pressure, followed by a decay over a period of up to a few tenths of a second (Figure 2.15a), the exact characteristics depending on the size and range of the explosion, and the topology of the region through which the blast wave travels. For a preliminary analysis, this may be treated as a very short impulse. Suppose that a moderate-sized explosion produces an impulse on a nearby low-rise building of 15 kNs. If the building is approximated by an SDOF system having dynamic mass 2.5 tonnes, natural period 0.2 s and 5% damping, estimate its response to the blast.

The natural period $\omega_n = 2\pi/0.2 = 31.42$ rad/s and $\omega_d = \omega_n\sqrt{(1-0.05^2)} = 31.38$ rad/s. Values can then simply be substituted into Equation 2.46 giving:

$$x = 0.191e^{-1.571t}\sin 31.38t$$

The motion is plotted in Figure 2.15b, and comprises a decaying oscillation with a period very close to 0.2 s. The largest displacement occurs at the first peak, $t = 0.049$ s, and takes the value $x = 0.177$ m.

This calculation is based on assumed linear behaviour and results in very large motions. For instance, if the structure being modelled were a single-storey building of height 3 m, this displacement would correspond to a drift ratio (i.e. maximum displacement divided by storey height) of nearly 6%. Even a building designed to resist lateral loads would be at risk of collapse at such a drift level. In reality, therefore, it is likely that a blast load of this magnitude would cause substantial damage; in the process, the structural properties of stiffness and damping would be likely to change significantly, resulting in quite different behaviour. The analysis of non-linear dynamic response is introduced in Chapter 5.

2.4.6 Response to irregular dynamic loads

Many environmental loads are highly irregular and do not lend themselves to simple mathematical description, making their analysis rather more complex. These can be approached in several ways, but the most conceptually simple one makes use of the impulse response introduced in Section 2.4.5.

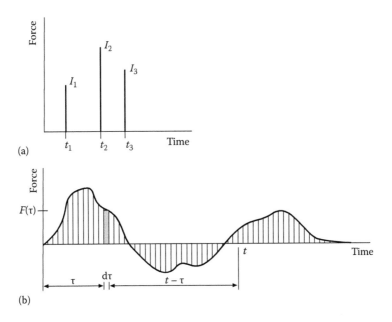

Figure 2.16 Treating irregular loads as sums of impulses: (a) three short pulses, (b) discretisation of a continuous load, leading to Duhamel's integral.

Consider the load shown in Figure 2.16a, made up of three discrete impulses. If we know the unit impulse response $h(t)$ of a dynamic system then, using linear superposition, the response to the impulse train at some time $t > t_3$ is simply the sum of the three individual impulse responses:

$$x(t) = I_1 h(t - t_1) + I_2 h(t - t_2) + I_3 h(t - t_3)$$

or, more generally, for a train of n impulses:

$$x(t) = \sum_{i=1}^{n} I_i h(t - t_i) \qquad (2.48)$$

Now consider the response to a continuous, irregular load (Figure 2.16b). By splitting the signal into very short time slices, we can represent this load as a long train of impulses. Take a typical slice at some time τ. Over the short duration $d\tau$, the force $F(\tau)$ varies very little, so the impulse is

$$I(\tau) = F(\tau) \cdot d\tau \qquad (2.49)$$

The displacement due to this impulse at some time $t > \tau$ is then

$$x(t) = I(\tau)h(t - \tau) \qquad (2.50)$$

But this is just the response to one tiny slice of the load. To find the total response at any given instant, we need to sum the responses to all the load slices up to that time, i.e. for all the possible values of τ between 0 and t. If we let $d\tau$ tend to zero, this results in an integral rather than a discrete summation:

$$x(t) = \int_0^t F(\tau)h(t-\tau)\,d\tau \tag{2.51}$$

Finally, we can substitute for the impulse response of a damped SDOF system from Equation 2.47, giving

$$x(t) = \frac{1}{m\omega_d} \int_0^t F(\tau)e^{-\xi\omega_n(t-\tau)}\sin\omega_d(t-\tau)\,d\tau \tag{2.52}$$

Equation 2.52 is known as Duhamel's integral. It is an example of a convolution integral, in which the product of two signals, one shifted in time relative to the other, is integrated over all possible values of the time shift. We will see more integrals of this type in Chapter 6, where we consider analysis methods that are suitable for random vibrations.

Bearing in mind that F is an irregular function of time, Duhamel's integral is not normally evaluated analytically, and some form of numerical integration must be used.

In the case of a ground motion, comparing the equations of motion 2.1 and 2.3, we replace absolute motion x by motion relative to the ground y, and applied force F by $-m\ddot{x}_g$, so that Duhamel's integral becomes

$$y(t) = \frac{-1}{\omega_d} \int_0^t \ddot{x}_g(\tau)e^{-\xi\omega_n(t-\tau)}\sin\omega_d(t-\tau)\,d\tau \tag{2.53}$$

Example 2.10

An undamped SDOF system has mass 800 kg and natural period 0.5 s. Calculate its response to the simple step load shown in Figure 2.17a. (Note that we would not normally use Duhamel's integral for such a simple load case as this – it could, for instance, be analysed as the sum of two step load responses, each calculated as in Section 2.4.1. The example is included here to illustrate the basic approach to evaluating a response by Duhamel's integral.)

The response is obtained by integrating all the impulse responses from time zero to time t using Equation 2.52. For values of t less than 0.1 s, there are no impulses, so the response is zero.

When t is between 0.1 and 0.5 s, then the load function, and hence the integral, can be split into two parts:

$$x(t) = \int_0^{0.1} 0.\,d\tau + \frac{1}{800.4\pi} \int_{0.1}^t 5000\sin 4\pi(t-\tau).\,d\tau$$

$$= \frac{5000}{3200\pi}\left[\frac{1}{4\pi}\cos 4\pi(t-\tau)\right]_{0.1}^t = \frac{25}{64\pi^2}[1-\cos 4\pi(t-0.1)]$$

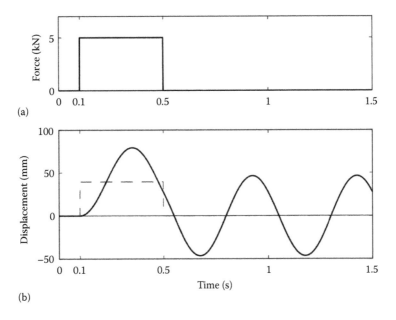

(a)

(b)

Figure 2.17 (a) Short-duration step load and (b) displacement response of undamped SDOF system as calculated by Duhamel's integral (static response shown dashed).

and when t is greater than 0.5 s, it can be split into three parts:

$$x(t) = \int_0^{0.1} 0.d\tau + \frac{1}{800.4\pi} \int_{0.1}^{0.5} 5000 \sin 4\pi(t-\tau).d\tau + \int_{0.5}^{t} 0.d\tau$$

$$= \frac{5000}{3200\pi} \left[\frac{1}{4\pi} \cos 4\pi(t-\tau) \right]_{0.1}^{0.5} = \frac{25}{64\pi^2} [\cos 4\pi(t-0.5) - \cos 4\pi(t-0.1)]$$

The response is plotted in Figure 2.17b. Note that there is still a response even after the loading has finished – this is a free vibration with the initial conditions obtained from the state of the system at the end of the loading. In a damped system, this would die away quite quickly, but damping has been omitted here to simplify the presentation of the method.

Now consider the response of damped systems to some more complex loads. The integrations quickly become lengthy and must be computed numerically, but the underlying behaviour is quite simple and can be illustrated with a simple example.

Figure 2.18 shows the responses of different 5% damped SDOF systems to a complex force time-history made up of numerous components in the range 1–6 Hz (or periods 1.0 down to 0.167 s). Plot (a) is the static displacement response that would be achieved if dynamic terms were ignored; this is obtained by simply dividing the force record by the stiffness. Plots (b)–(f) then show the responses of SDOF systems with ascending natural frequencies. In (b), the natural frequency of 0.5 Hz is below the lowest forcing frequency and as a result the system barely moves. In (c), (d) and (e), the natural frequency is in resonance with one of the frequency components of the loading, and so there is an amplified response at that frequency – while the response is not a pure sinusoid, it is clearly dominated by this single component. In (f), the natural frequency is above the range of the loading. In this case, the system responds quasi-statically, giving a response that is similar to the static one in (a).

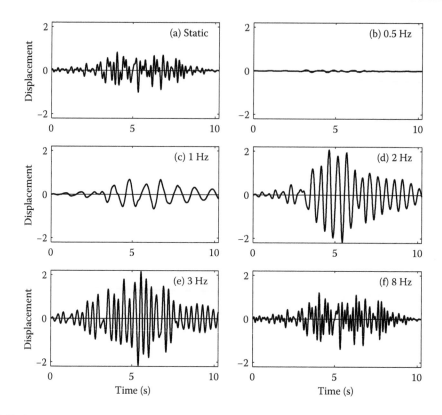

Figure 2.18 Time history responses of 5% damped SDOF systems computed by Duhamel's integral. The input force has frequency components evenly distributed in the range 1–6 Hz. (a) The static displacement F/k and (b)–(f) responses of oscillators with different frequencies. The displacement axes are normalised to give a peak static displacement of 1.

The SDOF system can be thought of as acting as a sort of filter-amplifier on the input force. Filters are widely used in communications and signal analysis to clean up signals. For instance, if a signal containing frequencies between, say, 5 and 10 kHz is transmitted from New York to London, it may pass through numerous cables, switches, wireless networks, etc. en route. The signal received in London will certainly not be identical to the one transmitted from New York because it will have been distorted by noise along the way. If we know that the New York signal was entirely concentrated between 5 and 10 kHz, then we also know that anything outside that range in the London signal is noise. We can improve the London signal by applying a band-pass filter which greatly attenuates the data outside the 5–10 kHz band, returning us to something close to the original signal – though not exactly, because no filter is perfect.

Thus, for example, the 2-Hz SDOF system in Figure 2.18d acts like a narrow band-pass filter and amplifier around that frequency. It amplifies the 2 Hz component and attenuates all the others, so that we are left with a signal mainly, but not entirely, at 2 Hz.

Example 2.11

An SDOF system has mass 2000 kg, stiffness 100 kN/m and damping of 5% of critical. Determine its displacement response to the 1940 El Centro earthquake ground motion.

The El Centro earthquake record was introduced in Chapter 1 (Figure 1.20) and, for convenience, is reproduced in Figure 2.19a. It is defined in the form of a record of horizontal

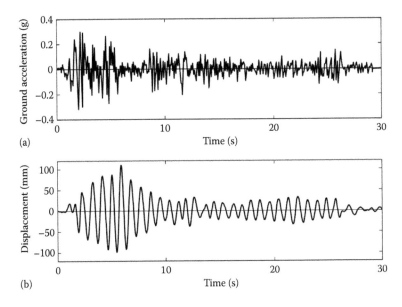

Figure 2.19 Response of a 5% damped SDOF system to the 1940 El Centro earthquake: (a) ground acceleration (g), (b) displacements (mm) computed by Duhamel's integral.

ground accelerations with time, digitised at intervals of 0.02 s. The motion is highly irregular, with numerous frequency components and a peak ground acceleration of 0.32 g.

For the SDOF system, we have seen that its responsiveness to a dynamic base motion
is dependent on the natural period and damping ratio, rather than on the absolute values
of mass, stiffness and damping. In this case, the natural period is

$$T_{n} = 2\pi\sqrt{\frac{m}{k}} = 2\pi\sqrt{\frac{2\times10^{3}}{10^{5}}} = 0.89 \text{ s}$$

This lies within the band of periods present in the ground motion, making a resonant
response likely. Given the complexity of the loading, explicit evaluation of Duhamel's
integral is not an option and we must resort to a numerical integration scheme. The
resulting displacement response is plotted in Figure 2.19b. Once again, we see that the
displacement of the mass is predominantly at the system's natural period – this component of the ground motion is picked up and amplified by the structure, with other components having relatively little effect on the dynamic behaviour.

2.4.7 Earthquake response spectrum

The response spectrum is a tool that is particularly used in earthquake engineering, though
its potential is wider. It is based on the idea that, in any dynamic loading event, we are not
that interested in how the displacements and forces within the structure vary over time – the
engineer only really wants to know what are the worst values that occur, as these are the
ones that will govern the design. This is what a response spectrum gives us; it provides a
summary of the largest response values of a linear SDOF system to a particular input loading time-history, as a function of the natural period and damping of the SDOF system.

The construction of a response spectrum can be visualised conceptually very simply.
Suppose the loading in which we are interested is the El Centro earthquake ground acceleration shown in Figure 1.20, Chapter 1.

For a given SDOF oscillator, defined by its natural period and damping ratio, we can compute the time history of the relative displacement between the mass and the ground using Equation 2.53. This is plotted for eight different systems (having four different natural periods and two levels of damping) in Figure 2.20. From each time history, the parameter we are really interested in is the *largest* relative displacement, which has been highlighted on each plot. It can be seen that the magnitude, sign and timing of this peak response vary with both the period and damping of the system. The magnitude (i.e. the biggest value, ignoring the sign) is called the *spectral displacement*, S_d; the sign has no real significance when considering a possible future earthquake whose directionality in relation to any particular building is unknown. The displacement response spectrum is then constructed by plotting the S_d values against the period of the SDOF oscillator, one curve for each damping level – this is shown at the bottom of Figure 2.20.

Of course, this illustrative spectrum is inadequate for practical use; we need to compute spectral values at much more closely spaced periods to get a representative curve. It also proves useful to plot spectra in terms of response quantities other than displacement. For instance, the *spectral acceleration*, S_a, is the magnitude (again ignoring the sign) of the absolute acceleration

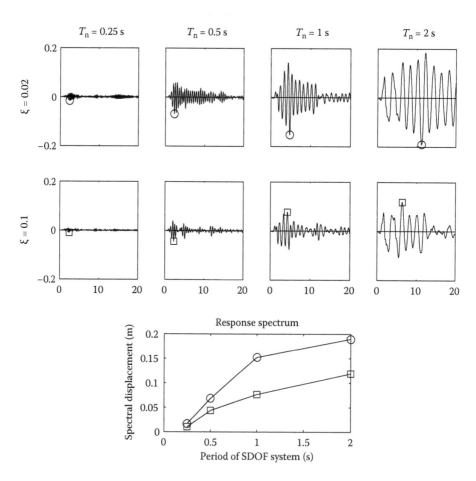

Figure 2.20 Construction of a response spectrum for the El Centro earthquake: the upper graphs plot displacement (m) versus time (s) for SDOF systems having four different natural periods and two damping ratios. The peak absolute values are then plotted against the natural period to form the spectrum in the bottom graph.

at the instant when $y = S_d$. Recall that the equation of motion for a system with a base acceleration, Equation 2.24, can be written in the form

$$(\ddot{x}_g + \ddot{y}) + 2\xi\omega_n\dot{y} + \omega_n^2 y = 0 \tag{2.54}$$

When the relative displacement takes its maximum value, $y = S_d$, the mass is at the end of its travel and so its velocity is zero. If the absolute acceleration at this instant is S_a, then Equation 2.54 becomes

$$S_a = \omega_n^2 S_d \tag{2.55}$$

It is also convenient to define a *spectral velocity*, S_v:

$$S_v = \omega_n S_d = \frac{S_a}{\omega_n} \tag{2.56}$$

Figure 2.21 shows acceleration, velocity and displacement spectra for the El Centro earthquake record, computed at closely spaced periods in the range 0–3 s, and at damping ratios

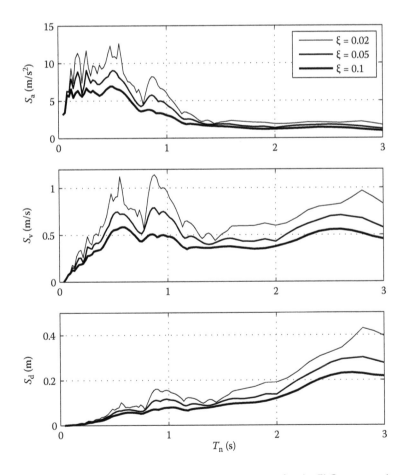

Figure 2.21 Acceleration, velocity and displacement response spectra for the El Centro earthquake at three damping levels.

of 2, 5 and 10%. These spectra contain a lot of information and can be confusing, so it is worth taking some time to think carefully about what they represent and how to use them. There are several important points to note here.

They are **response** spectra

Response spectra are not simply frequency domain representations of the earthquake time-history. Unlike the frequency spectrum calculated using the Fourier transform in Figure 1.21, the response spectra show the peak responses (in terms of acceleration, velocity or displacement) of a wide variety of SDOF systems to the earthquake time history. Nevertheless, there are some similarities, since both forms of spectra tell us something about how the energy of the earthquake record is distributed among its frequency components. In the case of the acceleration response spectrum, this is achieved because SDOF systems whose natural periods coincide with the dominant periods of the earthquake experience resonance, and so give higher spectral accelerations.

There are some similarities with harmonic response curves ...

The overall shapes of the acceleration response spectra are somewhat similar to the dynamic amplification curves due to a harmonic base motion, plotted in Figure 2.12. Both display the three regimes of response going from left to right: quasi-static, resonant and inertial.

At the left-hand end, the period of the SDOF system tends to zero, implying that it is effectively rigid. In this region, the mass simply moves in time with the ground and its acceleration is the same as the ground acceleration. Therefore, regardless of damping, the peak acceleration of the mass is equal to the peak ground acceleration, sometimes also called the *zero period acceleration* or ZPA. Therefore, the lines for different damping levels all converge to the same value of ZPA.

As we move to the right, we enter the region containing the predominant period components of the earthquake record. When applied to an SDOF system having a period in this range, a resonant amplification occurs, and so a large peak acceleration results. For the El Centro record, this occurs mainly in the range 0.2–0.6 s, with one or two further peaks up to around 1.0 s.

Beyond this period, the resonance effects reduce. Very long-period structures are extremely flexible and, to a first approximation, the mass stays still while the ground moves beneath it. They thus give quite large spectral velocities and displacements but very small accelerations.

For both the response spectrum and the harmonic response plots, increased damping reduces the dynamic amplification, giving flatter, less spiky curves.

... but also some important differences

In the acceleration response spectrum, rather than a single resonant peak, amplification occurs across the range of frequencies present in the ground motion. Crucially (and mercifully), the peak dynamic amplifications are much smaller – taking the 5% damped curve, the highest spectral accelerations are about a factor of three times the quasi-static value, whereas amplification factors close to 10 are achieved under harmonic loading.

The reduced response is due to the complexity and finite duration of the earthquake record. Whereas a harmonic base motion continues indefinitely at a single frequency, giving time for a resonant response to build up, in an earthquake the amplitude of a particular frequency component fluctuates throughout the loading, and ceases altogether after around 30 s, or often much less.

Spectral values are a mix of absolute and relative motions

Whereas S_d and S_v are defined in terms of *relative* motions, S_a is an *absolute* acceleration. This can seem confusing but perhaps makes more sense if we recall that, when multiplied through by the mass, Equation 2.54 is a statement of equilibrium between an inertia force (proportional to absolute acceleration) and damping and stiffness forces (proportional to relative velocity and displacement).

Spectral values are not always maximum values
While S_d is the largest relative displacement, S_a and S_v are only approximations to, respectively, the largest absolute acceleration and the largest relative velocity. We saw in Example 2.7 that it is possible to achieve a slightly larger acceleration a short time away from the maximum displacement point, when the relative velocity term is not zero. A similar result can be found for the velocity. However, the differences are not generally of great significance and will not be considered further here.

We do not usually need to plot all three spectra
The three forms of spectra are very simply related by Equation 2.56 – essentially, all three contain the same information, so two of them are superfluous. It is possible to plot a set of three acceleration, velocity and displacement spectra as a single curve on four-axis logarithmic graph paper, but most people find this confusing and rather difficult to read.

Because of its clear visual representation of resonance effects and its direct relationship with internal forces (discussed below), the acceleration spectrum is the most useful and is generally the only one plotted.

How to use response spectra
When analysing the behaviour of an SDOF structure under an earthquake load, we generally want to know the maximum force to which the structure will be subjected, and how far it moves. Given an acceleration response spectrum, if we know the natural period and damping of the structure, we can read the spectral acceleration S_a from the spectrum. The corresponding maximum displacement S_d can be found directly from Equation 2.55.

The inertia force corresponding to the spectral acceleration is mS_a. As stated above, at the instant when S_a and S_d occur, the relative velocity between the mass and the ground is zero, and hence so is the damping force. Therefore, this inertia force must be balanced by a stiffness force within the spring element of the SDOF system. This can also be seen by substituting for ω_n in Equation 2.55 to give

$$mS_a = kS_d \qquad\qquad (2.57)$$

Example 2.12

An idealised single-storey building has a mass (concentrated at roof level) of 1 tonne, a natural period of 0.2 s and 5% of critical damping. Estimate the force for which the columns must be designed and the maximum displacement of the roof caused by the El Centro (NS) earthquake record.

The spectral acceleration can simply be read. from Figure 2.21.

$S_a = 8.0$ m/s^2

The inertia force to be carried by the columns = $mS_a = 1000 \times 8.0 = 8000$ N = 8.0 kN

$\omega_n = 2\pi/T_n = 31.4$ rad/s

The maximum displacement of the roof $= \dfrac{S_a}{\omega_n^2} = \dfrac{8.0}{31.4^2} = 0.0081$ m = 8.1 mm

This roof displacement represents 0.25% of a typical storey height of 3.2 m, and is likely to be unproblematic.

By looking only at the acceleration spectrum, it can be tempting to think that matters could be improved by making the building more flexible, so that its period falls outside the range of large dynamic amplifications. Suppose (unrealistically) that we could increase the period to 1.0 s, the mass remaining unchanged. The spectral acceleration would reduce by a factor of nearly two, to 4.5 m/s^2, and therefore so would the inertia force, to 4.5 kN.

However, we now have $\omega_n = 2\pi/T_n = 6.28$ rad/s, and the maximum displacement of the roof becomes:

$$\frac{S_a}{\omega_n^2} = \frac{4.5}{6.28^2} = 0.113\,\text{m} = 113\,\text{mm}$$

Thus, we have paid for our reduced force demand with a 14-fold increase in the roof displacement, which is now 3.5% of a typical storey height, well above the normally accepted limit of 1%–2%. We must increase the stiffness of the building to reduce the deflections, and live with the resulting increased force demands. Of course, this problem can be seen at a glance by looking at the displacement spectrum.

The procedure above is an excellent way of determining the peak dynamic response characteristics of an SDOF system, but there is one further improvement we can make. The spectra in Figure 2.21 correspond only to one specific earthquake time history. If we were to compute spectra for a different time history, we would expect them to have a broadly similar shape, but with peaks occurring at slightly different periods. When considering what future earthquake might impact our structure, we cannot possibly know its exact form, but we can be reasonably confident its dominant periods will lie in the same range as other earthquakes that have occurred in the same or similar regions. Therefore, we can come up with a *design spectrum* by smoothing out and enveloping the spectra due to an appropriate set of earthquakes.

Figure 2.22 shows two examples of design spectra taken from Eurocode 8 (EC8), the European seismic design code of practice. Typically, codes will contain families of spectra, the appropriate choice depending on the nature of the earthquake and the local ground properties (since soils of different stiffness may amplify or attenuate certain period components). The examples shown are for what EC8 refers to as a Type I event (meaning a relatively large earthquake), with 5% structural damping. Spectra are shown for two ground types – A: rock, and C: stiff soil. In the latter case, the spectrum for the El Centro earthquake is also shown for comparison, and it can be seen that the EC8 spectrum does indeed provide a reasonable, smoothed curve which will envelope this and similar earthquakes as well.

2.5 TRANSFER FUNCTIONS AND BLOCK DIAGRAMS

An alternative analysis approach, widely used by mechanical and control engineers but underused by civil and structural engineers, is to represent a dynamic system as a *black box* or *block*, which takes an input (the applied loading or motion) and transforms it to give an output (e.g. the acceleration, velocity or displacement of a point within the structure). The transformation between input and output occurring within the block is defined by a transfer function, which is a simple multiplier applied to the input.

Unfortunately, it is not possible to define a system transfer function in the time domain – the time domain response of a linear dynamic system is given by the solution of a second

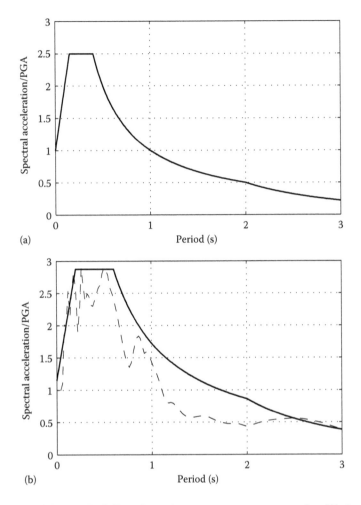

Figure 2.22 Examples of Eurocode 8 Type I acceleration response spectra for 5% damped structures. (a) Structures on ground type A – rock, (b) structures on ground type C – stiff soil, also showing 5% damped El Centro spectrum for comparison.

order differential equation and cannot be reduced to a simple multiplication. However, we can achieve the simplicity of the transfer function approach if we transform the problem into one in terms of a variable other than time.

2.5.1 Laplace transforms

In Chapter 1 we introduced the idea that a time-varying signal could instead be represented as a function of frequency, by breaking it down into its various frequency components using Fourier series (for periodic signals) or Fourier transforms (for non-periodic ones). A closely related transformation is the Laplace transform, which converts a time-varying signal into the Laplace domain, that is, into one in terms of the complex Laplace variable s. Some basic Laplace transform mathematics is presented in Appendix D, but a brief introduction will also be given here.

The Laplace transform $\bar{x}(s)$ of a time-varying signal $x(t)$ is defined as

$$\bar{x}(s) = \int_0^\infty e^{-st} x(t)\, dt \qquad (2.58)$$

where the Laplace variable s is a complex number. If $x(t)$ is a simple mathematical function, then evaluation of $\bar{x}(s)$ may be quite straightforward, but as $x(t)$ gets more complicated, so does the integration. This need not concern us, however, because Laplace transforms of most useful functions have been evaluated many times already, and can simply be looked up – a short table is provided in Appendix D.

A particularly important property of Laplace transforms is their relation to differentiation of the original, time-domain signal. Writing the Laplace transform as

$$x(t) \Rightarrow \bar{x}(s)$$

the first and second differentials transform according to

$$\dot{x}(t) \Rightarrow s\bar{x}(s) - x(0)$$

$$\ddot{x}(t) \Rightarrow s^2\bar{x}(s) - sx(0) - \dot{x}(0)$$

The value of this should immediately be obvious. Describing a dynamic system in the time domain normally requires two differentiations of $x(t)$ and the solution of a differential equation. However, in the Laplace domain, differentiation is replaced by a simple multiplication by the Laplace variable s, and the equation to be solved therefore becomes a much simpler polynomial.

Of course, it follows from the above that integration in the Laplace domain is achieved by dividing by s.

2.5.2 Transfer function of a linear SDOF system

We can now use Laplace transforms to derive the transfer function of a linear SDOF system. Figure 2.23 shows the standard representation of a mass-spring-damper system and its representation by a block, defined by a transfer function $H(s)$. We know that the time-domain equation of motion of this system is given by Equation 2.1:

$$m\ddot{x}(t) + c\dot{x}(t) + kx(t) = F(t)$$

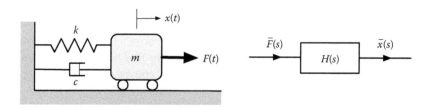

Figure 2.23 An SDOF system and the block diagram representation of its transfer function.

If we assume for the moment that all the initial conditions are zero, then taking the Laplace transform of this equation results in

$$ms^2\bar{x}(s) + cs\bar{x}(s) + k\bar{x}(s) = \bar{F}(s) \tag{2.59}$$

where $\bar{F}(s)$ is the Laplace transform of $F(t)$. Therefore, the transfer function, relating the output to the input is

$$H(s) = \frac{\bar{x}(s)}{\bar{F}(s)} = \frac{1}{ms^2 + cs + k} \tag{2.60}$$

or, using the normalised form of the system parameters:

$$H(s) = \frac{1}{m\left(s^2 + 2\xi\omega_n s + \omega_n^2\right)} \tag{2.61}$$

An alternative analysis approach now suggests itself. If we can find the Laplace transform of the input load, the response is easily found in the Laplace domain by multiplying by $H(s)$. The time domain response can then be found by taking the inverse Laplace transform – this can be the trickiest step, requiring some manipulation of $\bar{x}(s)$ to get it into an easily invertible form.

Example 2.13

In an experiment, a linear SDOF structure with mass m, damping coefficient c and stiffness k is shaken by an actuator, which is driven by an electrical signal $i(t)$ to provide a time-varying force $F(t) = Ai(t)$, where A is a constant. Draw a block diagram showing how the acceleration of the mass could be calculated in the Laplace domain.

The block diagram is shown in Figure 2.24. In each rectangular block, the output is simply obtained by multiplication of the input by the contents of the block. In the first block, the actuator takes an input i and multiplies it by a constant A to give the output force F. The transfer function of the structure can be obtained directly from Equation 2.60:

$$H(s) = \frac{1}{ms^2 + cs + k}$$

The second block applies this to the input force to give us the displacement response of the system. To obtain the velocity v and acceleration a, we must differentiate the displacement; in the Laplace domain this is achieved by multiplying by s.

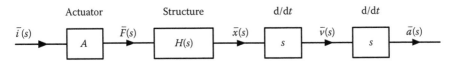

Figure 2.24 Block diagram representation of an SDOF structure loaded by an actuator.

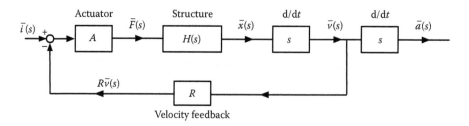

Figure 2.25 Modifying a block diagram by a velocity feedback loop.

Example 2.14

In the system of Example 2.13, it is found that the actuator is unable to provide a stable forcing when the velocity of the mass is high. To prevent this, it is decided to reduce the actuator force in proportion to the velocity. Modify the block diagram to show how this can be achieved.

This is an example of a *feedback control* system. Having calculated the velocity v of the mass we can scale it by some gain R. We then feed this back into the point in the block diagram where the actuator is being told how much force to apply (Figure 2.25). The circular junction is called a summing junction, its output being the sum of the inputs feeding into it.

2.5.3 Relationship to impulse response

In general terms, the equation of motion and its Laplace transform have the form

$$f(x(t) \text{ and its derivatives}) = F(t) \quad \Rightarrow \quad \bar{x}(s) = H(s)\bar{F}(s) \tag{2.62}$$

In the time domain, if the loading is a unit impulse, represented mathematically by the Dirac delta function $\delta(t)$, then the solution for $x(t)$ is called the impulse response and is normally denoted by $h(t)$. Now, as can be seen from Table D.1, the Laplace transform of $\delta(t)$ is simply 1, so Equation 2.62 becomes:

$$f(h(t) \text{ and its derivatives}) = \delta(t) \quad \Rightarrow \quad \bar{h}(s) = H(s).1 \tag{2.63}$$

Therefore, the transfer function of a linear system can be found by taking the Laplace transform of its unit impulse response. Both the impulse response (in the time domain) and the transfer function (in the Laplace domain) can be thought of as ways of defining fully the properties of a linear system, as alternatives to specifying $[m, c, k]$ or $[m, \xi, \omega_n]$.

This can be particularly useful in, for example, experimental work, where it may be difficult to measure m, c or k directly. In these circumstances, it may be far easier to measure a vibration response to an impulsive load (e.g. by hitting the structure with a hammer) and/ or a transfer function between a measured dynamic input (e.g. from a mechanical shaker) and the resulting output.

We will return to impulse response and transfer functions in Chapter 6, in the context of Fourier transforms.

KEY POINTS

- Using d'Alembert's principle, the equation of motion of an SDOF system can be conveniently visualised as an 'equilibrium' between the resultant external force and the internal stiffness, damping and inertia forces.
- The dynamic properties of a linear SDOF system can be represented in many different ways, all essentially containing the same information, for example: mass, damping and stiffness; mass, natural period and damping ratio; unit impulse response function; transfer function.
- For a linear system with constant mass, damping and stiffness, the equation of motion is a linear, second-order differential equation, which can be solved by standard mathematical techniques.
- It is always a good idea to start a dynamic analysis by solving the free vibration problem to find the natural frequency/period.
- The response to an oscillating dynamic load can be split into three regimes:
  - When the loading oscillates slowly compared to the structure's natural frequency, the response is quasi-static and stiffness-dominated;
  - When the load frequency is close to the structure's natural frequency, large dynamic amplifications occur due to resonance – the behaviour is damping-dominated;
  - When the load oscillates faster than the structure's natural frequency, the response is inertia-dominated and generally smaller than the static response.
- The severity of a dynamic response can be characterised by a dynamic amplification factor (DAF) – the ratio of the peak dynamic displacement to that achieved if a load of the same amplitude were applied statically.
- A harmonic load can result in extremely large DAFs (e.g. up to 25 with 2% damping). Environmental loads such as earthquakes tend to be irregular and of finite duration and so give lower DAFs.
- The dynamic response to irregular loads such as earthquakes can be evaluated using Duhamel's integral, a form of convolution integral.
- The response of a range of SDOF systems to an earthquake input is often summarised using a response spectrum, which plots the peak value of a response parameter (usually acceleration) as a function of the natural period and damping ratio.

TUTORIAL PROBLEMS

1. A light, simply supported beam of span 4 m has flexural rigidity $EI = 10^6$ Nm^2 and supports a mass of 1 tonne at its midspan. Find the natural frequency of the beam-mass system.

2. Figure P2.2 shows a solid cylinder of radius r, height h and density ρ_C floating upright in a fluid of density ρ_F. At its equilibrium position, a length L of the cylinder is submerged. Derive the equation of motion for small vertical oscillations about the equilibrium position, and hence find the natural frequency of the oscillations.

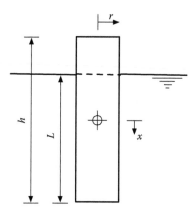

Figure P2.2

3. A mass oscillates back and forth along a circular surface as shown in Figure P2.3. Assuming that the mass slides rather than rolls (i.e. it has no rotational kinetic energy about its own centroid), use an energy approach to derive the equation of motion in terms of the angle θ. Show that, for *small* oscillations, this reduces to the standard linear form and find the natural frequency for this case.

4. A rigid, uniform bar of mass m and length L is supported by two springs of stiffness k and a central pivot as shown in Figure P2.4. It vibrates by rotating about the pivot. Derive the equation of motion in terms of the rotation angle θ. (Note: the moment of inertia of the bar about the pivot is $J = mL^2/12$.)

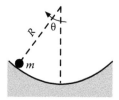

Figure P2.3

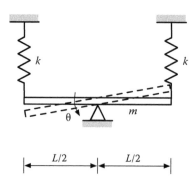

Figure P2.4

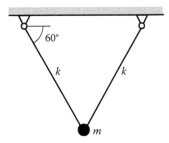

Figure P2.5

5. A mass m is supported by two pin-ended bars of axial stiffness k, in the form of an equilateral triangle (Figure P2.5). Find the natural frequency of vibrations if the mass moves: (a) purely vertically and (b) purely horizontally.

6. An SDOF mass-spring-damper system has a mass of 5 kg and stiffness 20 kN/m. If its damped natural frequency is measured as 8.0 Hz, calculate the damping ratio (i.e. the damping expressed as a fraction of the critical value).

7. How long will it take a 1-Hz free vibration to reduce to 5% of its initial value if its damping ratio is: (a) $\xi = 0.01$, (b) $\xi = 0.05$, (c) $\xi = 0.2$?

8. A set of scales consists of an instrumented spring from which the mass of the measured object can be deduced, and a damper to reduce oscillations of the readings. The mass of the scales is small and the damper is chosen to give critical damping when a 10-kg object is weighed. If a mass of 20 kg is placed on the scales, what is the damping ratio and how many cycles of vibration will occur before the reading on the scales settles to within 1% of the correct value?

9. A damped SDOF system has mass 100 kg, natural frequency 5.0 Hz and damping ratio 0.05. It is loaded by a harmonic force of amplitude 200 N, and whose frequency may lie anywhere in the range 6–10 Hz. Determine the peak steady state displacement and acceleration experienced by the mass.

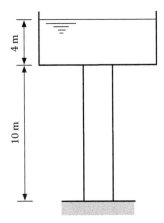

Figure P2.11

10. A vibration-sensitive instrument of mass 250 kg is positioned on a stiff floor slab. It is found that, in the worst case, the floor vibrates vertically at a frequency of 20 Hz and an amplitude of 1 mm. If the maximum allowable steady-state vibration amplitude of the instrument is 0.2 mm, determine the required stiffness of the spring mount between the floor and the instrument. Explain why adding damping does not help here.

11. Figure P2.11 shows an idealised water tower. The mass of the structure is 10 tonnes, mostly in the tank, and the capacity of the tank is 50 m³. When full of water, it has a natural period of lateral vibration of 1.0 s. It is situated in an area of high earthquake risk; the seismic load is defined by the smoothed response spectrum in Figure 2.22b, with a design peak horizontal ground acceleration of 4 m/s². Estimate the base shear force and moment and the maximum deflection of the tank when the water tower is: (a) full of water and (b) empty. (Neglect the fact that the water will slosh as the tank shakes. Take the centre of mass to be at the centre of the tank when full, and at the base of the tank when empty.)

Chapter 3

Multi-degree-of-freedom systems

3.1 INTRODUCTION

While single-degree-of-freedom (SDOF) systems provide an excellent way of introducing the important aspects of dynamic behaviour, and are reasonable representations of some structures, many real structures are complex and need to be modelled in greater detail. While in practice this is likely to be done using a finite element computer program, for visualisation purposes it is easiest to think of them as assemblages of lumped masses connected by members providing stiffness and damping.

Most of the basic principles governing the dynamic behaviour of single- and multi-degree-of-freedom (MDOF) systems are the same. There are two important differences. First, because there are now many motions to consider, the formulation involves vectors of nodal forces, accelerations, velocities and displacements, related to each other via matrices describing the structure's mass, damping and stiffness distributions. Analysis therefore requires some familiarity with matrix algebra. Secondly, while an SDOF system can only move in one way, a structure with n degrees of freedom can vibrate in any one (or many) of n different deformation shapes, or *modes*, each with its own natural frequency.

The basic structure of this chapter will follow that used for SDOF systems. First, we will formulate the equations of motion, then consider the free vibration characteristics, then use these as tools to help analyse the response to a forced vibration. When it comes to considering forced vibration response, we will first look at the response of 2DOF systems before moving on to the more general MDOF case. For clarity of presentation, many of the illustrations and examples will focus on systems with a small number of degrees of freedom, though the same methods and principles apply to larger systems. This chapter should enable you to

- Formulate equations of motion for MDOF systems,
- Understand how to derive natural frequencies and mode shapes for MDOF structures,
- Use the properties of mode shapes to simplify the equations governing MDOF vibrations,
- Analyse vibrations of undamped 2DOF systems by hand,
- Appreciate the basic principles of tuned mass dampers (TMDs),
- Formulate a linear dynamic analysis of an MDOF system using the mode superposition method,

- Formulate an appropriate representation of structural damping for a mode superposition analysis,
- Use a response spectrum in the context of a mode superposition analysis of an MDOF structure under an earthquake load.

3.2 EQUATIONS OF MOTION

3.2.1 MDOF systems with dynamic forces

Start with a simple 2DOF mass-spring-damper system, in which each mass is subjected to a separate, time-varying force. Figure 3.1 shows the full system, and the free body diagrams for each of the masses as they move. Note that, while the forces in spring and damper 1 depend only on the movement of mass 1, those for spring and damper 2 are related to the *relative* motion across the elements, that is, the difference in motion between mass 2 and mass 1. Applying Newton's second law to each mass in turn gives:

$$F_1 + c_2(\dot{x}_2 - \dot{x}_1) + k_2(x_2 - x_1) - c_1\dot{x}_1 - k_1 x_1 = m_1\ddot{x}_1$$
$$F_2 - c_2(\dot{x}_2 - \dot{x}_1) - k_2(x_2 - x_1) = m_2\ddot{x}_2 \tag{3.1}$$

We can write these more neatly by rearranging into the standard form of equation of motion and combining into a single matrix equation:

$$\begin{bmatrix} m_1 & 0 \\ 0 & m_2 \end{bmatrix}\begin{bmatrix} \ddot{x}_1 \\ \ddot{x}_2 \end{bmatrix} + \begin{bmatrix} c_1 + c_2 & -c_2 \\ -c_2 & c_2 \end{bmatrix}\begin{bmatrix} \dot{x}_1 \\ \dot{x}_2 \end{bmatrix} + \begin{bmatrix} k_1 + k_2 & -k_2 \\ -k_2 & k_2 \end{bmatrix}\begin{bmatrix} x_1 \\ x_2 \end{bmatrix} = \begin{bmatrix} F_1 \\ F_2 \end{bmatrix} \tag{3.2}$$

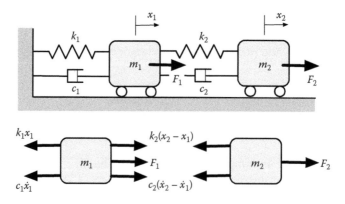

Figure 3.1 Two-degree-of-freedom mass-spring-damper system.

Using a bold, upright typeface to denote matrices and vectors, we can express this more compactly as

$$\mathbf{m\ddot{x}} + \mathbf{c\dot{x}} + \mathbf{kx} = \mathbf{f} \tag{3.3}$$

where $\mathbf{m}$ is the mass matrix, $\mathbf{c}$ is the damping matrix, $\mathbf{k}$ is the stiffness matrix, $\mathbf{f}$ is the load vector and $\mathbf{x}$ is the displacement vector. Equation 3.3 is a valid equation of motion for any MDOF system subjected to external forces, though obviously the size and complexity of the matrices may be much greater than in the example here.

Consider a larger example, of a multi-storey, *shear-type* building (Figure 3.2). Here, all the mass is assumed to be concentrated in the floors, which are effectively rigid. The columns provide lateral stiffness and some dashpots have been added to represent the damping within the structure – for now, don't worry about how to assign values to the dashpot coefficients. The floors are assumed to displace horizontally, giving the structure a shearing deformation. Figure 3.2 shows the free body diagram of a typical floor somewhere in the middle of the building. It experiences horizontal forces from the columns and dampers above and below it, proportional to the relative motions between it and the adjacent floors. Its equation of motion is therefore:

$$F_i + c_{i+1}(\dot{x}_{i+1} - \dot{x}_i) + k_{i+1}(x_{i+1} - x_i) - c_i(\dot{x}_i - \dot{x}_{i-1}) - k_i(x_i - x_{i-1}) = m_i\ddot{x}_i$$

or

$$m_i\ddot{x}_i - c_i\dot{x}_{i-1} + (c_i + c_{i+1})\dot{x}_i - c_{i+1}\dot{x}_{i+1} - k_ix_{i-1} + (k_i + k_{i+1})x_i - k_{i+1}x_{i+1} = F_i \tag{3.4}$$

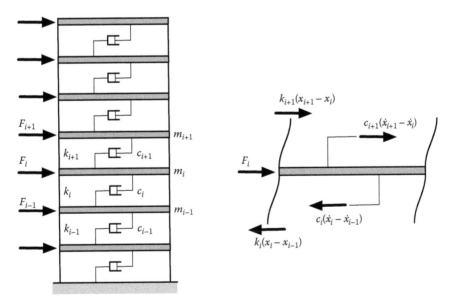

Figure 3.2 Setting up the equations of motion for a multi-storey shear-type building: general layout and free body diagram for a typical floor.

With many DOFs giving similar equations, the matrix form of Equation 3.3 looks like:

$$
\begin{bmatrix} * & & & & & & & \\ & * & & & & & \\ & & * & & & & \\ & & & * & & & \\ & & & & * & & \\ & & & & & * & \\ & & & & & & * \end{bmatrix}
\begin{bmatrix} \ddot{x}_1 \\ \cdots \\ \ddot{x}_{i-1} \\ \ddot{x}_i \\ \ddot{x}_{i+1} \\ \cdots \\ \ddot{x}_n \end{bmatrix}
+
\begin{bmatrix} * & * & & & & & \\ * & * & * & & & & \\ & * & * & * & & & \\ & & * & * & * & & \\ & & & * & * & * & \\ & & & & * & * & * \\ & & & & & * & * \end{bmatrix}
\begin{bmatrix} \dot{x}_1 \\ \cdots \\ \dot{x}_{i-1} \\ \dot{x}_i \\ \dot{x}_{i+1} \\ \cdots \\ \dot{x}_n \end{bmatrix}
$$

$$
+
\begin{bmatrix} * & * & & & & & \\ * & * & * & & & & \\ & * & * & * & & & \\ & & * & * & * & & \\ & & & * & * & * & \\ & & & & * & * & * \\ & & & & & * & * \end{bmatrix}
\begin{bmatrix} x_1 \\ \cdots \\ x_{i-1} \\ x_i \\ x_{i+1} \\ \cdots \\ x_n \end{bmatrix}
=
\begin{bmatrix} F_1 \\ \cdots \\ F_{i-1} \\ F_i \\ F_{i+1} \\ \cdots \\ F_n \end{bmatrix}
\qquad (3.5)
$$

where each * denotes some non-zero value and the blank matrix cells are all zero. Equation 3.5 is a system of n coupled differential equations. The nature of this set of equations, and why they are hard to solve, is discussed further in Section 3.2.3.

Example 3.1

Figure 3.3a shows a very simplified 2D idealisation of a bridge, in which the deck mass is lumped at three points ($2m$ each) and the mass of the pier is lumped at its centre (m). Restraint to vertical movement is provided by the flexural stiffness k of the deck segments and the much higher axial stiffness, $10k$, of the pier segments. Neglecting damping, write the equations of motion for vertical vibrations in matrix form.

Assume the masses undergo vertical displacements $x_1 - x_4$, where DOFs 1, 2 and 3 are in the deck and DOF 4 is in the column. The resulting spring forces on each mass are shown in Figure 3.3b. Resolving vertically for each mass:

$$F_1 + k(x_2 - x_1) - kx_1 = 2m\ddot{x}_1$$

$$F_2 + k(x_3 - x_2) - k(x_2 - x_1) - 10k(x_2 - x_4) = 2m\ddot{x}_2$$

$$F_3 - k(x_3 - x_2) - kx_3 = 2m\ddot{x}_3$$

$$10k(x_2 - x_4) - 10kx_4 = m\ddot{x}_4$$

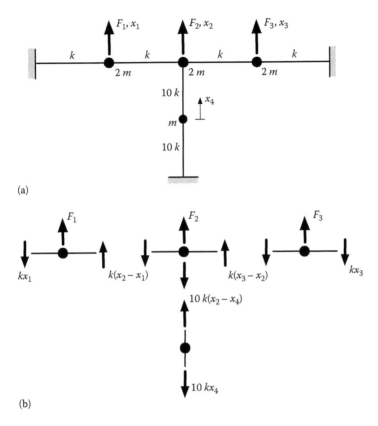

Figure 3.3 Vibration of two-span bridge: (a) 4DOF idealisation of the structure and (b) free body diagrams for each DOF showing internal forces due to the displacements $x_1 - x_4$.

which can be written in matrix form as:

$$
m\begin{bmatrix} 2 & 0 & 0 & 0 \\ 0 & 2 & 0 & 0 \\ 0 & 0 & 2 & 0 \\ 0 & 0 & 0 & 1 \end{bmatrix}\begin{bmatrix} \ddot{x}_1 \\ \ddot{x}_2 \\ \ddot{x}_3 \\ \ddot{x}_4 \end{bmatrix} + k\begin{bmatrix} 2 & -1 & 0 & 0 \\ -1 & 12 & -1 & -10 \\ 0 & -1 & 2 & 0 \\ 0 & -10 & 0 & 20 \end{bmatrix}\begin{bmatrix} x_1 \\ x_2 \\ x_3 \\ x_4 \end{bmatrix} = \begin{bmatrix} F_1 \\ F_2 \\ F_3 \\ 0 \end{bmatrix}
$$

Note: it is worth emphasising how greatly simplified this model is. In practice, the structure would deform horizontally as well as vertically, giving more degrees of freedom and hence larger matrices. We also haven't thought about how the mass and stiffness values could be calculated and how the mass should be lumped to give a realistic representation of the dynamic response, though these issues were discussed in Chapter 1.

3.2.2 MDOF systems subjected to base motions

Suppose the shear-type building introduced above were subjected to a base motion $x_g(t)$ instead of dynamic forces (Figure 3.4). As with the SDOF case, we can define the motion

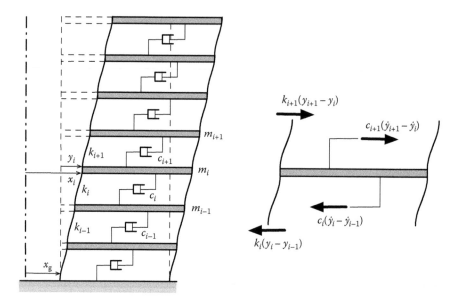

Figure 3.4 Multi-storey shear-type building with a base motion: general layout and free body diagram for a typical floor.

of the ith DOF either as an absolute displacement x_i, or as a displacement y_i relative to the ground:

$$y_i = x_i - x_g \tag{3.6}$$

The equation of motion for this DOF can then be written as

$$c_{i+1}(\dot{y}_{i+1} - \dot{y}_i) + k_{i+1}(y_{i+1} - y_i) - c_i(\dot{y}_i - \dot{y}_{i-1}) - k_i(y_i - y_{i-1}) = m_i(\ddot{y}_i + \ddot{x}_g)$$

or

$$m_i\ddot{y}_i - c_i\dot{y}_{i-1} + (c_i + c_{i+1})\dot{y}_i - c_{i+1}\dot{y}_{i+1} - k_iy_{i-1} + (k_i + k_{i+1})y_i - k_{i+1}y_{i+1} = -m_i\ddot{x}_g \tag{3.7}$$

The left-hand side of Equation 3.7 is directly comparable to that of Equation 3.4, replacing x by y. The right-hand side shows that the ground motion is equivalent to applying a force to each DOF equal to its mass times the ground acceleration. The equations can be assembled into matrix form as before, and written compactly as:

$$\boldsymbol{m\ddot{y} + c\dot{y} + ky = -m\mathbf{i}\ddot{x}_g} \tag{3.8}$$

i is a vector of influence coefficients, which allows us to specify which DOFs are excited by the ground motion. For instance, in a 3D model, DOFs in the direction of the ground motion would be excited and would be assigned influence coefficients of one, while those perpendicular to it would not, and would be given influence coefficients of zero. If all DOFs are excited, then **i** is simply a vector of ones and the product **mi** is a column vector of all the nodal masses.

3.2.3 Some comments on MDOF equations of motion

From the cases presented, it should already be possible to notice some important features of the matrix equations governing the dynamics of MDOF systems.

First, and most obviously, if an analytical model of a system has n degrees of freedom, then each vector in Equation 3.3 has n elements, and each matrix is $n \times n$. Although a modern computer can solve even large matrix equations quite quickly, it nevertheless makes sense to try to avoid making n too large. Excessive model complexity can be confusing for the analyst, for whom managing and interpreting the output become significant issues.

For a linear elastic system, the mass, damping and stiffness matrices *must* all be symmetric, that is, they are unchanged by writing their rows as columns (generally referred to as *transposing*). Terms on the leading diagonal are always positive, and off-diagonal terms are usually negative or zero.

The stiffness matrix is exactly the same as would be used for a static analysis. It is a banded matrix, with non-zero elements on the leading diagonal and other nearby diagonals, but mostly zeros elsewhere. For a simple series of masses and springs like the shear-type building, **k** is tri-diagonal (i.e. a bandwidth of 3). A finite element model will generally use a more realistic discretisation of a structure, resulting in many DOFs and a somewhat higher bandwidth.

In all the examples shown, a very straightforward approach has been taken to describing the mass distribution within the structure. It is simply represented by an appropriately chosen point mass at each DOF. This results in a *lumped* mass matrix, which has non-zero elements only on the leading diagonal. This is a perfectly valid approach, though it has the slight drawback that it can require quite a large number of DOFs (possibly more than would be used in a static analysis) in order to model the mass distribution reasonably accurately. Some finite element programs offer an alternative formulation, known as a *consistent* mass matrix, in which the nodal masses are derived by integration of the finite element shape functions. This allows the mass distribution to be more accurately modelled using fewer DOFs, at the expense of introducing off-diagonal terms into the mass matrix. Examples in this book will all assume lumped mass, though the general principles of the solution methods are the same in either case.

For an SDOF system, we saw that it can be difficult to define the damping coefficient c. Instead, the equations are generally expressed using a non-dimensionalised damping parameter such as the damping ratio ξ or logarithmic decrement δ. For an MDOF system, this difficulty is multiplied; to model the damping explicitly would require estimating the coefficients of numerous viscous dashpots distributed across the structure. For this reason, the damping matrix is not generally formulated in the manner shown above. Instead, damping is incorporated in the analysis by another method, such as modal damping, in which each mode of vibration is assigned a damping ratio, or Rayleigh damping, in which the damping matrix is taken as a linear combination of the mass and stiffness matrices. These will be explained more fully later.

Lastly, it is worth emphasising that we have, rather easily, created a very complex mathematical problem. Equations 3.3 and 3.8 are n-dimensional systems of coupled, second-order, ordinary differential equations (where 'coupled' means that we do not have a single ODE for each DOF – each line of the matrix equation includes terms relating to several adjacent DOFs). We will see that we can greatly simplify the solution by first solving the free vibration problem, then expressing the governing equations in terms of the free vibration characteristics.

3.3 FREE VIBRATION OF MDOF SYSTEMS

3.3.1 Natural frequencies and mode shapes

Before considering the response to an applied load, we solve the free vibration problem, with the load set to zero. We showed in the case of SDOF systems that damping causes free vibrations to decay over time, and that it causes a change in the vibration frequency. However, for practical levels of damping, the latter effect is small. We will therefore neglect damping for the time being, so that Equation 3.3 reduces to

$$m\ddot{x} + kx = 0 \tag{3.9}$$

The displacement $\mathbf{x}$ varies as a function of both position (i.e. DOF) and time. We can express this as the product of a function of position only, say $\mathbf{u}$, and a function of time only. For SDOF systems, the variation of displacement with time was sinusoidal, and it is reasonable to assume it will take a similar form in the MDOF case. So we can write

$$\mathbf{x} = \mathbf{u} \sin \omega t \tag{3.10}$$

Since $\mathbf{u}$ is not a function of time, differentiating Equation 3.10 involves only differentiation of the sine term. Then substituting into Equation 3.9 gives

$$(\mathbf{k} - \omega^2 \mathbf{m})\mathbf{u} \sin \omega t = 0 \tag{3.11}$$

This is sometimes known as the characteristic equation of the system. There are three possible ways in which Equation 3.11 can be true. Either $\mathbf{u}$ or $\sin \omega t$ could be zero, but both of these solutions imply that the system is not vibrating. This leaves only the third option, which is that the matrix $(\mathbf{k} - \omega^2 \mathbf{m})$ has no inverse. This in turn requires that its determinant is zero. We can therefore solve:

$$\det(\mathbf{k} - \omega^2 \mathbf{m}) = 0 \tag{3.12}$$

to find values of natural frequency ω. For a system with n DOFs, Equation 3.12 is an nth-order polynomial that can be solved to give n values of ω (not necessarily all different). Each natural frequency in turn can then be substituted into

$$(\mathbf{k} - \omega^2 \mathbf{m})\mathbf{u} = 0 \tag{3.13}$$

to find a corresponding vector $\mathbf{u}$. Thus, an n-DOF system has n natural frequencies ω_i, each of which is associated with a corresponding deformation shape $\mathbf{u}_i$ called the *mode shape*.

As we will see, the solution of Equation 3.13 does not allow $\mathbf{u}_i$ to be determined fully. It gives us ratios between the elements of $\mathbf{u}_i$, with the overall amplitude undefined. As the structure vibrates at a particular natural frequency, the overall magnitude of the motion varies with time, but the ratios between the magnitudes of the various DOFs stays the same, as defined by the mode shape. The amplitude of vibration is determined by the initial conditions in a free vibration, and by the external load in the forced case. In the absence of either, we must apply some arbitrary scaling to the mode shape. A commonly used approach is *mass normalisation*, in which each modal vector is scaled in such a way that

$$\mathbf{u}_i^{\mathrm{T}} \mathbf{m} \mathbf{u}_i = 1 \tag{3.14}$$

The mathematical significance of the above analysis can be seen if we multiply Equation 3.13 by the inverse of the mass matrix:

$$(\mathbf{m}^{-1}\mathbf{k} - \omega^2\,\mathbf{I})\mathbf{u} = 0 \tag{3.15}$$

where $\mathbf{I}$ is an $n \times n$ unit matrix. This is in the form of a standard equation for the eigenvalues λ and eigenvectors $\mathbf{u}$ of a matrix $\mathbf{A}$:

$$(\mathbf{A} - \lambda\mathbf{I})\mathbf{u} = 0 \tag{3.16}$$

Comparing Equations 3.15 and 3.16, we see that the circular natural frequencies of an MDOF system are the square roots of the eigenvalues of $\mathbf{m}^{-1}\mathbf{k}$, and the mode shapes are the corresponding eigenvectors. This is important because: (a) there exist numerous very efficient numerical methods for determining eigenvalues of large matrices (for which the determinant approach is impractical), and (b) the eigenvectors have some useful properties which can help us simplify the matrix equations. The basic theory of eigenvalues and vectors is presented in Appendix E.

In most of the examples in this chapter, we will use mass-normalised mode shapes because of their analytical convenience. However, other forms of normalisation are possible and are equally valid. For example, it can be convenient to normalise modes so that their largest element is 1, or so that their value at a particular DOF is always 1.

Example 3.2

Figure 3.5a shows a small-scale laboratory model of a 2-storey shear-type building where $k = 2500$ N/m and $m = 25$ kg. Find the natural frequencies and mode shapes.

Using the approach outlined in Section 3.2.1, the equations of motion are

$$\begin{bmatrix} 2m & 0 \\ 0 & m \end{bmatrix}\begin{bmatrix} \ddot{x}_1 \\ \ddot{x}_2 \end{bmatrix} + \begin{bmatrix} 2k & -k \\ -k & k \end{bmatrix}\begin{bmatrix} x_1 \\ x_2 \end{bmatrix} = 0$$

so

$$\mathbf{k} - \omega^2\mathbf{m} = \begin{bmatrix} 2(k - \omega^2 m) & -k \\ -k & k - \omega^2 m \end{bmatrix}$$

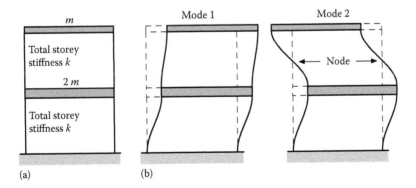

Figure 3.5 (a) Two-storey shear-type frame and (b) its mode shapes.

and

$$\det(\mathbf{k} - \omega^2\mathbf{m}) = 2(k - \omega^2 m)(k - \omega^2 m) - k^2 = 2\omega^4 m^2 - 4\omega^2 km + k^2 = 0$$

This is a quadratic equation for ω^2. Noting that $k/m = 100\ \text{s}^{-2}$, this can be solved to give

$$\omega_1^2 = \frac{k}{m}\left(1 - \frac{1}{\sqrt{2}}\right) \rightarrow \omega_1 = 0.541\sqrt{\frac{k}{m}} = 5.41\ \text{rad/s} \rightarrow f_1 = 0.86\ \text{Hz}$$

$$\omega_2^2 = \frac{k}{m}\left(1 + \frac{1}{\sqrt{2}}\right) \rightarrow \omega_2 = 1.307\sqrt{\frac{k}{m}} = 13.07\ \text{rad/s} \rightarrow f_2 = 2.08\ \text{Hz}$$

To find the first mode shape, substitute ω_1 into Equation 3.13 to give

$$k\begin{bmatrix} \sqrt{2} & -1 \\ -1 & 1/\sqrt{2} \end{bmatrix}\begin{bmatrix} u_1 \\ u_2 \end{bmatrix} = 0 \rightarrow \sqrt{2}u_1 = u_2$$

Note that both lines of the matrix give the same relationship between u_1 and u_2. The mass normalised mode shape can be found by guessing some arbitrary scaling, calculating the resulting value of the product $\mathbf{u}^T\mathbf{m}\mathbf{u}$ and adjusting accordingly. In this case we get

$$u_1 = \frac{1}{2\sqrt{m}}\begin{bmatrix} 1 \\ \sqrt{2} \end{bmatrix} = \frac{1}{10}\begin{bmatrix} 1 \\ \sqrt{2} \end{bmatrix} = \begin{bmatrix} 0.1 \\ 0.141 \end{bmatrix}$$

You might like to check that this does indeed give $\mathbf{u}^T\mathbf{m}\mathbf{u} = 1$. Finally, following an identical process using ω_2 gives

$$u_2 = \frac{1}{2\sqrt{m}}\begin{bmatrix} 1 \\ -\sqrt{2} \end{bmatrix} = \frac{1}{10}\begin{bmatrix} 1 \\ -\sqrt{2} \end{bmatrix} = \begin{bmatrix} 0.1 \\ -0.141 \end{bmatrix}$$

The mode shapes are sketched (with the displacements shown in correct proportion) in Figure 3.5b. As is generally the case, the lower-frequency mode involves a simpler deformed shape, which requires less strain energy for a given amplitude than the more complex second mode. In the first mode, both storeys oscillate back and forth in phase, while the higher mode involves anti-phase motion: one storey moves right while the other moves left, in a continually alternating sequence. Note also that, in the second mode, there is a point around the mid-height of the second storey which has zero displacement. This point, called the *node*, remains stationary as points above and below it vibrate in opposite directions. In general, we expect the mode shape of mode i to contain $(i - 1)$ nodes.

Example 3.3

Find the natural frequencies and mode shapes of the mass-spring system in Figure 3.6, in which the mass unit $m = 1$ kg and the stiffness $k = 40$ N/m.

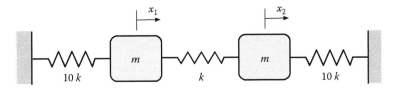

Figure 3.6 Weakly coupled 2DOF system.

Setting up the equations of motion in the usual way gives

$$m \begin{bmatrix} 1 & 0 \\ 0 & 1 \end{bmatrix} \begin{bmatrix} \ddot{x}_1 \\ \ddot{x}_2 \end{bmatrix} + k \begin{bmatrix} 11 & -1 \\ -1 & 11 \end{bmatrix} \begin{bmatrix} x_1 \\ x_2 \end{bmatrix} = 0$$

(You can check this by drawing the free body diagrams yourself.) Insert the usual trial solution:

$$\begin{bmatrix} x_1 \\ x_2 \end{bmatrix} = \begin{bmatrix} u_1 \\ u_2 \end{bmatrix} \sin \omega t$$

leading to:

$$\begin{bmatrix} 11k - \omega^2 m & -k \\ -k & 11k - \omega^2 m \end{bmatrix} \begin{bmatrix} u_1 \\ u_2 \end{bmatrix} \sin \omega t = 0$$

The natural frequencies are found by setting the determinant of the matrix to zero, and the mode shapes by back-substituting the frequencies into the above equation. Inserting the numerical values for k and m and mass-normalising the mode shapes, we get

$$\omega_1 = 20 \, \text{rad/s} \; (f_1 = 3.2 \, \text{Hz}) \rightarrow u_1 = \frac{1}{\sqrt{2}} \begin{bmatrix} 1 \\ 1 \end{bmatrix}$$

$$\omega_2 = 21.9 \, \text{rad/s} \; (f_2 = 3.5 \, \text{Hz}) \rightarrow u_2 = \frac{1}{\sqrt{2}} \begin{bmatrix} 1 \\ -1 \end{bmatrix}$$

This is an example of a *weakly coupled* system. It can be visualised as two SDOF systems (the outer mass-spring systems), each having natural frequency $\omega_n = \sqrt{(10k/m)} = 20$ rad/s, coupled by the flexible, central spring. In the first mode, the two SDOF systems oscillate in phase and with equal amplitudes, so the central spring remains undeformed. It therefore contributes nothing to the stiffness of this mode, and the resulting frequency is the same as if the central spring were not there. In the second mode, the SDOF systems are in anti-phase and so the coupling spring comes into play. However, because the spring has comparatively low stiffness, the frequency of this second mode is only slightly higher than that of the two SDOF systems.

Example 3.4

In the 2DOF system of Example 3.3, mass 1 is given an initial displacement of 20 mm with mass 2 held stationary, and then both masses are released. Plot their subsequent displacement time histories.

In general, setting an MDOF system in motion and allowing it to vibrate freely will result in simultaneous vibrations of each of its natural modes. We therefore expect a motion that is a linear combination of the vibration in each mode, with the modal amplitudes governed by the initial conditions. Because we are imposing an initial displacement, we would expect the motion to follow a cosine rather than a sine function – remember that a cosine is just a sine wave with a time-shifted origin. The motion can therefore be written as

$$\begin{bmatrix} x_1 \\ x_2 \end{bmatrix} = \frac{1}{\sqrt{2}} \begin{bmatrix} 1 \\ 1 \end{bmatrix} A \cos\omega_1 t + \frac{1}{\sqrt{2}} \begin{bmatrix} 1 \\ -1 \end{bmatrix} B \cos\omega_2 t$$

where A and B are the modal amplitudes, to be determined from the initial conditions. Using units of mm, at time $t = 0$ this reduces to

$$\begin{bmatrix} 20 \\ 0 \end{bmatrix} = \frac{1}{\sqrt{2}} \begin{bmatrix} 1 \\ 1 \end{bmatrix} A + \frac{1}{\sqrt{2}} \begin{bmatrix} 1 \\ -1 \end{bmatrix} B$$

from which $A = B = 10\sqrt{2}$.

Therefore, the two displacements are

$$x_1 = 10(\cos\omega_1 t + \cos\omega_2 t)$$

$$x_2 = 10(\cos\omega_1 t - \cos\omega_2 t)$$

These are plotted in Figure 3.7. Each vibration looks like the product of two sine or cosine functions, one having quite a high frequency and the other a low frequency (long period), which has the effect of modulating the amplitude of the motion. It is also noticeable that the motions are out of phase, so that the amplitude of one DOF is largest when the other is zero, and *vice versa*. These features can be demonstrated mathematically by using simple trigonometric identities to rewrite the displacement expressions as

$$x_1 = 20\cos\frac{(\omega_1 + \omega_2)t}{2}.\cos\frac{(\omega_1 - \omega_2)t}{2}$$

$$x_2 = -20\sin\frac{(\omega_1 + \omega_2)t}{2}.\sin\frac{(\omega_1 - \omega_2)t}{2}$$

So the higher frequency is the mean of the two natural frequencies (about 21 rad/s, or 3.35 Hz in this case) and the lower one is half the difference (about 1 rad/s or 0.15 Hz). The phenomenon evident in Figure 3.7, in which the energy (and therefore the vibration amplitude) flows continually back and forth between the two masses at a lower frequency than the main vibration frequency, is known as *beating*.

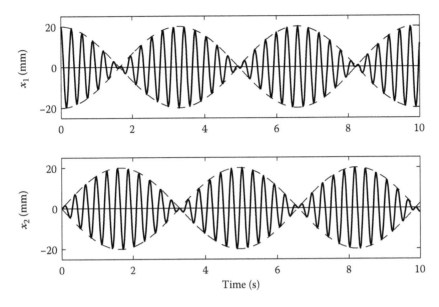

Figure 3.7 Free vibrations resulting from an initial displacement of mass 1 in the weakly coupled 2DOF system of Figure 3.6, showing the phenomenon of beating.

Example 3.5

Draw the mode shapes for the bridge whose equations of motion were derived in Example 3.1.

This structure has four DOFs, making manual solution of the matrix equations imprac-tical, but they can easily be solved computationally. The values, together with plots of the mode shapes, are shown in Figure 3.8, in order of ascending natural frequency, and can be described as follows:

- In mode 1 all DOFs move in phase – both deck spans flex upward equally as the column extends, then flex downward as the column contracts.
- Mode 2 comprises the deck spans bending in anti-phase, with the column stationary.
- In mode 3 the deck spans bend in phase with each other, but out of phase with the column – as the deck flexes upward, the column shortens, and vice versa.

$$f_1 = 0.143\sqrt{\frac{k}{m}} \qquad f_2 = 0.159\sqrt{\frac{k}{m}} \qquad f_3 = 0.287\sqrt{\frac{k}{m}} \qquad f_4 = 0.763\sqrt{\frac{k}{m}}$$

$$\mathbf{u}_1 = \begin{bmatrix} 0.48 \\ 0.19 \\ 0.48 \\ 0.10 \end{bmatrix}\frac{1}{\sqrt{m}} \qquad \mathbf{u}_2 = \begin{bmatrix} 0.50 \\ 0 \\ -0.50 \\ 0 \end{bmatrix}\frac{1}{\sqrt{m}} \qquad \mathbf{u}_3 = \begin{bmatrix} 0.14 \\ -0.63 \\ 0.14 \\ -0.37 \end{bmatrix}\frac{1}{\sqrt{m}} \qquad \mathbf{u}_4 = \begin{bmatrix} -0.01 \\ 0.27 \\ -0.01 \\ -0.92 \end{bmatrix}\frac{1}{\sqrt{m}}$$

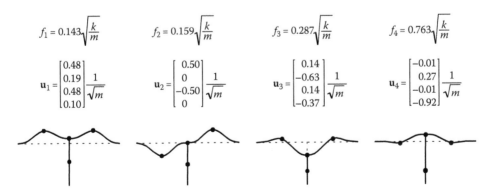

Figure 3.8 Natural frequencies and mode shapes for the 4DOF bridge model.

- In mode 4 the two lumped masses in the column move in anti-phase, implying extension of the top part of the column as the bottom compresses, and vice versa. While such a mode is physically possible, its presence here is really an artefact of the rather crude way in which the mass of the column has been lumped.

Again, it is evident that the higher frequency modes require the development of more strain energy in the structure – more complex deformed shapes, more anti-phase motion, more deformation of the stiffer elements.

3.3.2 Properties of mode shapes

The fact that they are eigenvectors of the mass and stiffness matrices gives the mode shapes some special mathematical properties that will be useful when analysing the response of an MDOF system to an external force. The underlying maths is explained a little more fully in Appendix E.

First, it can be shown that any deformed shape $\mathbf{x}$ of a structure can be written as a linear combination of its mode shapes, that is:

$$\mathbf{x} = X_1\mathbf{u}_1 + X_2\mathbf{u}_2 + X_3\mathbf{u}_3 \ldots = \mathbf{UX} \tag{3.17}$$

where $\mathbf{U}$ is a matrix comprising the mode shapes written as column vectors and $\mathbf{X}$ is a column vector of the modal amplitudes, often called *modal displacements* or *generalised displacements*. In fact, the modes prove particularly efficient at doing this, so that it is often possible to get a good approximation to the deformed shape of a structure with many DOFs using only its first few mode shapes.

Secondly, the mode shapes, being eigenvectors, have important *orthogonality* properties in relation to the mass and stiffness matrices, meaning that if we pre- and post-multiply either $\mathbf{m}$ or $\mathbf{k}$ by two *different* mode shape vectors $\mathbf{u}_i$ and $\mathbf{u}_j$, the result is zero.

$$\mathbf{u}_i^\mathrm{T}\mathbf{m}\mathbf{u}_j = \mathbf{u}_i^\mathrm{T}\mathbf{k}\mathbf{u}_j = 0 \quad i \neq j \tag{3.18}$$

With the modal vectors mass normalised according to Equation 3.14, we know that pre- and post-multiplying the mass matrix by the *same* mode shape $\mathbf{u}_i$ gives the answer 1. If we pre-multiply Equation 3.13 by $\mathbf{u}_i^\mathrm{T}$ and rearrange slightly, we get

$$\mathbf{u}_i^\mathrm{T}\mathbf{k}\mathbf{u}_i = \omega_i^2\mathbf{u}_i^\mathrm{T}\mathbf{m}\mathbf{u}_i = \omega_i^2 \tag{3.19}$$

This can all be summarised quite concisely by pre- and post-multiplying by the matrix of mode shapes $\mathbf{U}$ rather than by one mode shape at a time:

$$\mathbf{U}^\mathrm{T}\mathbf{m}\mathbf{U} = \mathbf{I}, \quad \mathbf{U}^\mathrm{T}\mathbf{k}\mathbf{U} = \mathbf{\Lambda} \quad \text{where } \mathbf{\Lambda} = \begin{bmatrix} \omega_1^2 & 0 & 0 & 0 \\ 0 & \omega_2^2 & 0 & 0 \\ 0 & 0 & \ddots & 0 \\ 0 & 0 & 0 & \omega_n^2 \end{bmatrix} \tag{3.20}$$

So the mode shapes can be used to *diagonalise* the mass and stiffness matrices, which will greatly simplify the solution of the equations of motion. We will see how this works in Section 3.5.

Example 3.6

For the 2DOF structure of Example 3.2, show that the following deflected shapes can be expressed as a linear combination of the mode shapes:

$$\begin{bmatrix} x_1 \\ x_2 \end{bmatrix} = \begin{bmatrix} 1 \\ 1 \end{bmatrix}, \begin{bmatrix} 1 \\ 2 \end{bmatrix}, \begin{bmatrix} 1 \\ 0 \end{bmatrix}$$

From Equation 3.17, the vector **x** can be expressed in terms of the mode shape matrix and the modal amplitude coefficients X_1 and X_2 as follows:

$$\begin{bmatrix} x_1 \\ x_2 \end{bmatrix} = \frac{1}{10} \begin{bmatrix} 1 & 1 \\ \sqrt{2} & -\sqrt{2} \end{bmatrix} \begin{bmatrix} X_1 \\ X_2 \end{bmatrix}$$

This can be inverted to give

$$\begin{bmatrix} X_1 \\ X_2 \end{bmatrix} = 5 \begin{bmatrix} 1 & 1/\sqrt{2} \\ 1 & -1/\sqrt{2} \end{bmatrix} \begin{bmatrix} x_1 \\ x_2 \end{bmatrix}$$

Then, substituting in each of the deflected shape vectors in turn gives

$$\begin{bmatrix} x_1 \\ x_2 \end{bmatrix} = \begin{bmatrix} 1 \\ 1 \end{bmatrix} \rightarrow \begin{bmatrix} X_1 \\ X_2 \end{bmatrix} = \begin{bmatrix} 8.54 \\ 1.46 \end{bmatrix}, \quad \begin{bmatrix} x_1 \\ x_2 \end{bmatrix} = \begin{bmatrix} 1 \\ 2 \end{bmatrix} \rightarrow \begin{bmatrix} X_1 \\ X_2 \end{bmatrix} = \begin{bmatrix} 12.07 \\ -2.07 \end{bmatrix},$$

$$\begin{bmatrix} x_1 \\ x_2 \end{bmatrix} = \begin{bmatrix} 1 \\ 0 \end{bmatrix} \rightarrow \begin{bmatrix} X_1 \\ X_2 \end{bmatrix} = \begin{bmatrix} 5 \\ 5 \end{bmatrix}$$

Thus, the first and second deformed shapes involve motion predominantly in mode 1, while the third entails equal contributions from each mode.

Example 3.7

For the bridge problem introduced in Example 3.1, check that the mode shapes given in Example 3.3 do indeed diagonalise the stiffness matrix as set out in Equation 3.20.

Matrix multiplication gives

$$\mathbf{U}^{\mathsf{T}} \mathbf{k} \mathbf{U} = \frac{k}{m} \begin{bmatrix} 0.48 & 0.19 & 0.48 & 0.10 \\ 0.50 & 0 & -0.50 & 0 \\ 0.14 & -0.63 & 0.14 & -0.37 \\ -0.01 & 0.27 & -0.01 & -0.92 \end{bmatrix}$$

$$\begin{bmatrix} 2 & -1 & 0 & 0 \\ -1 & 12 & -1 & -10 \\ 0 & -1 & 2 & 0 \\ 0 & -10 & 0 & 20 \end{bmatrix} \begin{bmatrix} 0.48 & 0.50 & 0.14 & -0.01 \\ 0.19 & 0 & -0.63 & 0.27 \\ 0.48 & -0.50 & 0.14 & -0.01 \\ 0.10 & 0 & -0.37 & -0.92 \end{bmatrix}$$

$$= \frac{k}{m} \begin{bmatrix} 0.807 & 0 & 0 & 0 \\ 0 & 1.0 & 0 & 0 \\ 0 & 0 & 3.24 & 0 \\ 0 & 0 & 0 & 22.95 \end{bmatrix}$$

Alternatively, formulating the matrix Λ from the natural frequencies gives

$$\Lambda = (2\pi)^2 \frac{k}{m} \begin{bmatrix} 0.143^2 & 0 & 0 & 0 \\ 0 & 0.159^2 & 0 & 0 \\ 0 & 0 & 0.287^2 & 0 \\ 0 & 0 & 0 & 0.763^2 \end{bmatrix}$$

$$= \frac{k}{m} \begin{bmatrix} 0.807 & 0 & 0 & 0 \\ 0 & 1.0 & 0 & 0 \\ 0 & 0 & 3.24 & 0 \\ 0 & 0 & 0 & 22.95 \end{bmatrix}$$

The results are the same, showing that Equation 3.20 is satisfied.

3.3.3 Empirical methods

In preliminary calculations and/or simplified methods of design, it is often desired to estimate the fundamental modal properties of a structure without creating a detailed computer model. For multi-storey buildings, a variety of formulae have been proposed. The simplest and crudest rule of thumb is that each storey of a building represents a period of 0.1 s, so that the fundamental period (in s) or frequency (in Hz) of an n-storey building can be estimated from

$$T_1 = \frac{n}{10}; \quad f_1 = \frac{10}{n} \tag{3.21}$$

More recent formulae have acquired a little more sophistication, recognising the differences in stiffness across structural systems and construction materials. For example, the formulae adopted for regular buildings in the European seismic code, Eurocode 8, relate the fundamental period (in s) to the building height h (in m):

$$T_1 = C_t h^{0.75} \tag{3.22}$$

where the coefficient C_t varies with construction type:

- Steel moment-resisting frames (i.e. where resistance to lateral loads comes mainly from the moment resistance of the beam-column joints): $C_t = 0.085$
- Concrete moment-resisting frames: $C_t = 0.075$
- Steel or concrete concentrically braced frames (diagonal braces pass through beam-column joints): $C_t = 0.05$
- Steel eccentrically braced frames (braces offset from beam-column joints): $C_t = 0.075$

Figure 3.9 shows a comparison between Equations 3.21 and 3.22 for buildings up to 20 storeys, assuming a typical storey height of 3 m. On the basis of this comparison, Equation 3.21 seems to give a reasonable approximation to the fundamental period for the more flexible frame types.

While they may look very crude, there is no particular reason to consider period estimates obtained by these simple formulae as any less reliable as those achieved using a full analysis. Although entirely empirical, Equation 3.22 has been calibrated against a large database of measurements from full-scale buildings, complete with real ground conditions, cladding, contents, and so on, aspects that can have a significant influence on dynamic properties but are generally dealt with simplistically or not at all in an analytical model.

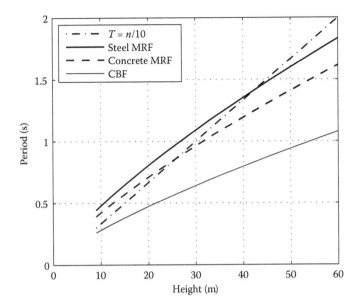

Figure 3.9 Comparison of approximate formulae for fundamental period of multi-storey buildings, assuming a storey height of 3 m.

3.4 VIBRATION ANALYSIS OF 2DOF SYSTEMS

Before moving to the more general MDOF case, we will take a look at some aspects of the vibration of 2DOF systems. For undamped 2DOF systems, it is possible to set up and solve the equations of motion directly; though even for this relatively simple case, the algebra gets quite long-winded. We will take a brief look at this because it can provide some insight into MDOF system behaviour. However, it has limited practical application. A more generally applicable model of an MDOF system will normally have more degrees of freedom, and will include damping. In this case, the modal superposition approach (Section 3.5.1) is generally the best way to proceed.

3.4.1 Equivalence between 2DOF and two SDOF systems

An undamped 2DOF system can be represented by two mass-spring systems in series, and so can be thought of as two SDOF systems glued together. Suppose we have SDOF systems with properties $[m_A, k_A]$ and $[m_B, k_B]$. Each system has a natural frequency defined by

$$\omega_A^2 = \frac{k_A}{m_A}, \quad \omega_B^2 = \frac{k_B}{m_B} \tag{3.23}$$

If we now put the two systems together to create a single 2DOF system as shown in Figure 3.10, how do the natural frequencies of this new system relate to those of the constituent SDOF systems? The equation of motion for free vibrations is

$$\begin{bmatrix} m_A & 0 \\ 0 & m_B \end{bmatrix} \begin{bmatrix} \ddot{x}_A \\ \ddot{x}_B \end{bmatrix} + \begin{bmatrix} k_A + k_B & -k_B \\ -k_B & k_B \end{bmatrix} \begin{bmatrix} x_A \\ x_B \end{bmatrix} = 0 \tag{3.24}$$

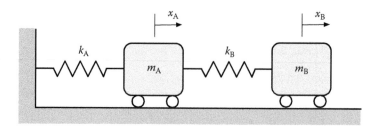

Figure 3.10 Undamped 2DOF system created by combining SDOF systems A and B.

The natural frequencies ω_1 and ω_2 are found by putting det $(\mathbf{k} - \omega^2\mathbf{m}) = 0$. Substituting for m_A, m_B, k_A and k_B from Equation 3.23, the resulting expressions can be written as

$$\omega_1^2 = \frac{1}{2}\left[\omega_A^2 + (1+\mu)\omega_B^2 - \left(\omega_A^4 + (1+\mu)^2\omega_B^4 + 2(\mu-1)\omega_A^2\omega_B^2\right)^{1/2}\right]$$

$$\omega_2^2 = \frac{1}{2}\left[\omega_A^2 + (1+\mu)\omega_B^2 + \left(\omega_A^4 + (1+\mu)^2\omega_B^4 + 2(\mu-1)\omega_A^2\omega_B^2\right)^{1/2}\right]$$

(3.25)

where the mass ratio μ is

$$\mu = \frac{m_B}{m_A}$$

(3.26)

Equation 3.25 is plotted in Figure 3.11, each graph corresponding to a different ratio between the SDOF frequencies ω_A and ω_B. All three cases show similar trends. The 2DOF system always has one natural frequency lower than the lower of the two SDOF frequencies (in which the two masses move in phase) and one higher than the higher of the two SDOF frequencies (in which the masses move in anti-phase). When the second mass is small (μ close

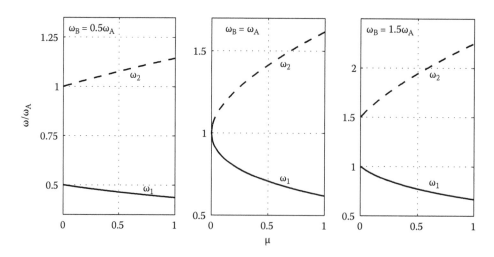

Figure 3.11 Natural frequencies of 2DOF systems in terms of the frequencies of their SDOF components.

to zero), the interaction between the two SDOF systems is negligible, and the frequencies of the combined system tend toward those of the two separate systems. As the mass ratio increases, so does the difference between the SDOF frequencies and those of the combined system.

Example 3.8

In vibration problems involving humans, there is a growing recognition that the human body is a dynamic system in its own right, and not simply a rigid mass that sits on a structure. A simple illustration of this is the dynamics of a floor supporting a standing person. Consider a floor with a first natural frequency of 6.8 Hz, supporting a person or people with a first vertical natural frequency of 5.0 Hz (Figure 3.12a). Although both the floor and the person are complex dynamic systems, we will assume here that the behaviour of each can be reasonably approximated by an SDOF system (Figure 3.12). The ratio of the mass of the people to the floor is 0.12.

Referring to the preceding analysis, let the floor be system A and the person system B. Then we have $\omega_A = (2\pi) \times 6.8$ rad/s, $\omega_B = (2\pi) \times 5$ rad/s and $\mu = 0.12$. Equation 3.25 gives $f_1 = 4.71$ Hz and $f_2 = 7.21$ Hz.

It is also instructive to calculate the mode shapes associated with each frequency. This can be done by substituting the frequencies back into Equation 3.13. In this case, the mode shapes have been arbitrarily normalised so that the largest element has a magnitude of unity.

$$\mathbf{u}_1 = \begin{bmatrix} 0.111 \\ 1 \end{bmatrix}, \quad \mathbf{u}_2 = \begin{bmatrix} -1 \\ 0.925 \end{bmatrix}$$

The modes are sketched in Figure 3.12c, with the modal displacements drawn in the correct proportions. The first mode, which has a frequency slightly lower than that of the person, involves in-phase vibration of the person and the floor. However, the floor motion is small and the mode is dominated by human body vibration. The second mode involves roughly equal-amplitude, anti-phase vibrations of the floor and body, at a frequency a little higher than that of the floor on its own.

Note: This is, of course, a highly simplified example. Many models of the human body involve two or more DOFs, and the inherent damping within the body can be around 30% of critical, so that it cannot reasonably be neglected.

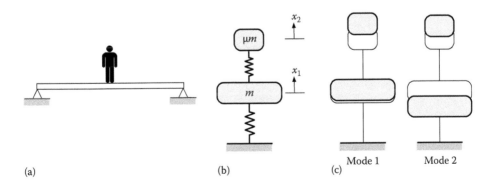

(a)   (b)   (c)

Figure 3.12 (a) Person-floor system, (b) its 2DOF idealisation and (c) the mode shapes.

3.4.2 Undamped response to a harmonic load

Consider the undamped 2DOF mass-spring system in Figure 3.13, with a harmonic load applied to mass 1. The equations of motion are easy to derive:

$$\begin{bmatrix} m_1 & 0 \\ 0 & m_2 \end{bmatrix} \begin{bmatrix} \ddot{x}_1 \\ \ddot{x}_2 \end{bmatrix} + \begin{bmatrix} k_1 + k_2 & -k_2 \\ -k_2 & k_2 \end{bmatrix} \begin{bmatrix} x_1 \\ x_2 \end{bmatrix} = \begin{bmatrix} F\sin\omega_F t \\ 0 \end{bmatrix} \tag{3.27}$$

The natural frequencies of the system can be found by applying Equation 3.12, that is, putting $\det(\mathbf{k} - \omega^2\mathbf{m}) = 0$, giving

$$(k_1 + k_2 - \omega^2 m_1)(k_2 - \omega^2 m_2) - k_2^2 = 0 \tag{3.28}$$

This results in a quadratic equation that can be solved for the two natural frequencies ω_1 and ω_2 and the corresponding mode shapes can then be found (if required) by substituting each frequency in turn into Equation 3.13.

We can then consider the response to the forcing. Similarly to an undamped SDOF system (Section 2.4.2), we would expect the steady state response to be of the form:

$$\begin{bmatrix} x_1 \\ x_2 \end{bmatrix} = \begin{bmatrix} X_1 \\ X_2 \end{bmatrix} \sin\omega_F t \tag{3.29}$$

Differentiating and substituting into the equations of motion gives

$$\begin{bmatrix} k_1 + k_2 - m_1\omega_F^2 & -k_2 \\ -k_2 & k_2 - m_2\omega_F^2 \end{bmatrix} \begin{bmatrix} X_1 \\ X_2 \end{bmatrix} = \begin{bmatrix} F \\ 0 \end{bmatrix} \tag{3.30}$$

This matrix equation can be inverted by standard methods, though the algebra gets quite complicated. The resulting amplitudes of the two masses are

$$\begin{bmatrix} X_1 \\ X_2 \end{bmatrix} = \begin{bmatrix} k_2 - m_2\omega_F^2 \\ k_2 \end{bmatrix} \frac{F}{m_1 m_2 \left(\omega_1^2 - \omega_F^2\right)\left(\omega_2^2 - \omega_F^2\right)} \tag{3.31}$$

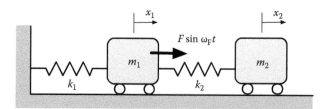

Figure 3.13 Undamped, harmonically loaded 2DOF system.

Example 3.9

For the 2DOF system shown in Figure 3.14a, show how the displacement amplitudes of the masses vary with the forcing frequency ω_F.

In this case, $k_1 = k_2 = k$ and $m_1 = m_2 = m$ so that Equation 3.28 becomes

$$(2k - \omega^2 m)(k - \omega^2 m) - k^2 = 0$$

from which the natural frequencies can be found as

$$\omega_1^2 = \frac{(3 - \sqrt{5})}{2}\frac{k}{m}, \quad \omega_2^2 = \frac{(3 + \sqrt{5})}{2}\frac{k}{m}$$

or

$$\omega_1 = 0.618\sqrt{\frac{k}{m}}, \quad \omega_2 = 1.618\sqrt{\frac{k}{m}}$$

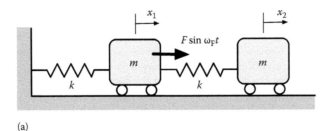

(a)

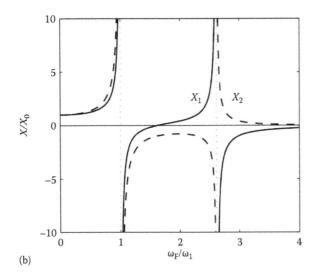

(b)

Figure 3.14 (a) 2DOF system with equal masses and stiffness subjected to harmonic load and (b) response amplitudes as a function of forcing frequency (DOF 1 – solid, DOF 2 – dashed).

The corresponding mode shapes are not really needed for the remainder of the problem, but it is good practice to calculate them and check they are reasonable. For simplicity, they are given below with (arbitrarily) the amplitude of DOF 1 normalised to unity:

$$\mathbf{u}_1 = \begin{bmatrix} 1 \\ (1+\sqrt{5})/2 \end{bmatrix} = \begin{bmatrix} 1 \\ 1.618 \end{bmatrix}, \quad \mathbf{u}_2 = \begin{bmatrix} 1 \\ (1-\sqrt{5})/2 \end{bmatrix} = \begin{bmatrix} 1 \\ -0.618 \end{bmatrix}$$

As expected, the first mode entails the two masses moving in phase and the second in anti-phase.

The amplitudes of displacement caused by the harmonic load are then given by Equation 3.31. If we define a static displacement $X_0 = F/k$, then these can be written as

$$\begin{bmatrix} X_1 \\ X_2 \end{bmatrix} = \begin{bmatrix} k - m\omega_F^2 \\ k \end{bmatrix} \frac{kX_0}{m^2\left(\omega_1^2 - \omega_F^2\right)\left(\omega_2^2 - \omega_F^2\right)}$$

The variation of the storey response amplitudes with forcing frequency is plotted in Figure 3.14b. This looks complicated but in fact is quite straightforward. There are two resonances, at the natural frequencies ω_1 and ω_2, indicated by vertical dotted lines. Below the first resonance, the motion of each mass is similar, with X_2 somewhat larger. Around the first resonance, X_1 and X_2 have the same sign and around the second resonance they take opposite signs, as would be expected from the mode shapes at these frequencies. There is a transition point at $\omega_F = 1.618\omega_1 = \sqrt{k/m}$, where the motion of mass 1 is zero. At this point, the external force exerted on mass 1 is exactly cancelled out by the force from the spring connecting it to mass 2, so mass 1 does not move.

3.4.3 Tuned mass dampers

When problematic vibrations arise in structures, there is often a need to find some way of reducing them by adding damping to the system. One way of doing this is to add discrete viscous dampers to the structure to increase the effective damping ratio. An alternative approach is to use a *tuned mass damper* (TMD) – a secondary mass-spring system that counter-acts the vibrations due to the applied loads.

Suppose a structure can be represented as an SDOF system with properties m_A, k_A giving a natural frequency $\omega_A = \sqrt{k_A/m_A}$, subjected to a harmonic load $F(t) = F \sin \omega_F t$. If ω_F is close to ω_A, then unacceptable vibrations may arise due to resonance. To reduce these, we can add a secondary mass-spring system with natural frequency $\omega_B = \sqrt{k_B/m_B}$ where

$$\omega_B = \omega_A \rightarrow k_B = m_B\omega_A^2 \tag{3.32}$$

The natural frequencies ω_1 and ω_2 of the 2DOF system thus created can be found from Equation 3.25; one will be slightly lower and one slightly higher than ω_A and ω_B. Substituting into Equation 3.31 and simplifying gives

$$X_A = 0, \quad X_B = \frac{F}{k_B} \tag{3.33}$$

For convenience, we would like our damper to be reasonably small, with a mass ratio $\mu = m_B/m_A$ of the order of 5%–10%. However, the smaller we make m_B, the larger will be the amplitude X_B. The choice of the TMD properties therefore involves some compromise between its size and its amplitude of motion.

The above analysis ignores the effect of any viscous damping. In practice, the TMD could be improved by adding a viscous dashpot to the mass-spring system, removing energy from the structure and reducing the amplitude of the TMD motion. However, the analysis then becomes more complex and is not covered here.

As its name implies, to be effective, a TMD must be quite accurately tuned to the problematic vibration frequency. If vibrations occur over a range of frequencies, or if the problem frequency is liable to vary over time, then a TMD is unlikely to provide an effective solution.

Example 3.10

The fundamental mode of a footbridge is idealised as an SDOF system with natural frequency 2.5 Hz, mass 5 tonnes and very low damping. A group of people crossing the bridge is approximated as a harmonic load of amplitude 5 kN and frequency 2.5 Hz – this is a typical, comfortable pacing rate. Design a TMD to control the displacement of the deck, subject to a displacement limit of 50 mm. If, in reality, people's pacing frequency over the bridge varies in the range 2.0–3.0 Hz, estimate the peak displacement of the deck and the TMD mass over this range.

To control the vibrations, set the TMD frequency $\omega_B = 2\pi \times 2.5 = 15.7$ rad/s. Equation 3.33 then gives:

$$k_B = \frac{5000}{0.05} = 100\,\text{kN/m} \quad \text{and} \quad m_B = \frac{k_B}{\omega_B^2} = \frac{100 \times 10^3}{15.7^2} = 405\,\text{kg}$$

The mass ratio of the TMD is $\mu = 405/5000 = 0.081$. From Equation 3.25, the natural frequencies of the combined bridge-TMD system are $f_1 = 2.17$ Hz and $f_2 = 2.88$ Hz.

With the input frequency varying over a range, the amplitude of the response of the deck without the TMD can be estimated from the SDOF harmonic solution, Equation 2.30, approximating its inherent damping as zero. When the TMD is added, the amplitudes of the bridge deck and TMD can be found from Equation 3.31. Both solutions are plotted in Figure 3.15. With a forcing frequency of 2.5 Hz, the TMD performs as required, with the deck deflection reduced to zero and the TMD deflection limited to 50 mm. Between about 2.25 and 2.75 Hz, the deck continues to deflect less than in the absence of the TMD but at the expense of somewhat larger motions of the TMD itself. Outside this range the response becomes dominated by resonance with the modes of the combined deck-TMD system, with unacceptably large vibrations of both elements.

Clearly the TMD design is less than ideal in this case because it does not give a good solution over the full range of frequencies (though it is worth noting that walking frequencies at the extremes of this range, particularly above about 2.8 Hz, are quite unusual). In practice, the TMD could be improved by the incorporation of a viscous dashpot, and might form just one part of a more complex vibration control system. For example, in the retrofit of the London Millennium Bridge, TMDs were used to damp the most problematic vibration mode, supplemented by a series of discrete viscous dampers which were effective over a much broader frequency range. (See Section 1.1.1, Chapter 1 and the Further Reading section at the end of the book for more details.)

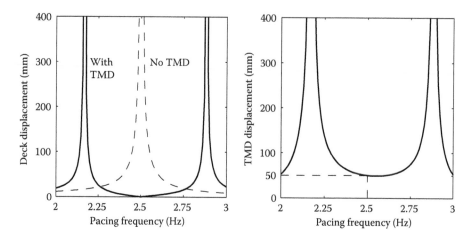

Figure 3.15 Controlling footbridge vibrations with a TMD: displacement amplitudes of deck (left) and TMD (right).

3.5 RESPONSE OF MDOF SYSTEMS TO DYNAMIC LOADS

While 2DOF systems can sometimes be analysed by hand, dynamic analysis of most MDOF systems leads to large, complex matrix equations for which an efficient computational solution method is required. Here we will focus on the mode superposition method, which takes advantage of the modal properties discussed in Section 3.3.2 to simplify the matrix equations for linear systems.

3.5.1 Mode superposition method: theory

We now wish to solve the full matrix equations of motion, with an external forcing vector f, Equation 3.3, with any number of degrees of freedom:

$$m\ddot{x} + c\dot{x} + kx = f$$

We saw in Equation 3.17 that it is possible to express the displacements in terms of the mode shapes:

$$x(i,t) = U(i)X(t)$$

Here, the displacements **x** are functions of both position (which we define in terms of DOF i) and time t. They are written as a combination of U, a matrix containing the mode shapes written as column vectors, which have arbitrary amplitude and are functions only of position, and a vector of time-varying amplitudes **X** known as *modal displacements*. Differentiating with respect to time requires only differentiation of the **X** terms, with the U terms acting as constant coefficients. Substitute into the equation of motion and then pre-multiply each term by U^T:

$$U^{T}mU\ddot{X} + U^{T}cU\dot{X} + U^{T}kUX = U^{T}f \qquad (3.34)$$

which can be written more compactly as

$$M\ddot{X} + C\dot{X} + KX = F \tag{3.35}$$

where the terms

$$M = U^T mU, \quad C = U^T cU, \quad K = U^T kU, \quad F = U^T f \tag{3.36}$$

are known as, respectively, the *generalised mass, damping and stiffness matrices*, and the *generalised force vector*.

We saw in Section 3.3.2 that both M and K are diagonal matrices. As discussed earlier, the original damping matrix c is difficult to formulate explicitly, so it is hard to know whether its generalised form C is diagonal. While there is no particular reason to suppose that it would be diagonal, we will nevertheless make that assumption for the time being, and think later about how to make it so. Each line of Equation 3.36 then represents an independent differential equation in terms of one of the elements of the vector X:

$$M_i\ddot{X}_i + C_i\dot{X}_i + K_iX_i = F_i \tag{3.37}$$

or, dividing through by M_i:

$$\ddot{X}_i + \frac{C_i}{M_i}\dot{X}_i + \omega_i^2 X_i = \frac{F_i}{M_i} \tag{3.38}$$

Thus, by converting the equations of motion from ones in terms of the actual displacements x to ones in terms of the modal amplitudes X, we have succeeded in removing the coupling between them, making them much easier to solve. The equation of motion for each modal displacement is just like an SDOF equation of motion, and can be solved using the methods introduced in Chapter 2, once we have decided how to find C_i. Once the X_i are known, the actual displacements can then be found from Equation 3.17.

Finally, note that if we are using mass-normalised mode shapes, then $M_i = 1$ for all i, and Equation 3.38 can be simplified accordingly. In the full matrix form, M becomes a unit matrix and $K = \Lambda$, a diagonal matrix of the squared circular natural frequencies, so that Equation 3.35 can then be written as

$$\ddot{X} + C\dot{X} + \Lambda X = F \tag{3.39}$$

3.5.2 Modelling damping in MDOF systems

Several approaches have been developed to formulate a damping matrix that both provides a reasonably accurate estimate of the damping and satisfies the modal orthogonality condition, that is, can be diagonalised by the mode shapes. Two of the simpler and more widely used ones are modal damping and Rayleigh damping.

3.5.2.1 Modal damping

Looking at Equation 3.38, it is tempting to write

$$\frac{C_i}{M_i} = 2\xi_i\omega_i \tag{3.40}$$

so that Equation 3.38 becomes

$$\ddot{X}_i + 2\xi_i\omega_i\dot{X}_i + \omega_i^2 X_i = \frac{F_i}{M_i} \tag{3.41}$$

which is exactly analogous to the acceleration-units equation of motion for an SDOF system, Equation 2.23. ξ_i is called the *modal damping ratio*, and defines the damping level in a particular mode of vibration in the same way as ξ does for an SDOF system. With this approach, the damping can be very easily incorporated into the modal equations of motion without ever having to calculate the original damping matrix **c** or perform the computations needed to convert it into its modal form.

Using modal damping allows us, if we wish, to specify a different damping ratio for each of the structure's modes of vibration. However, a reasonable default option is simply to specify the same damping ratio in all modes. For SDOF structures, we argued that a simple, global damping measure is sufficient to give an appropriate level of accuracy. A similar argument can be made for MDOF structures; if we are happy to use the approximation of, say, 5% of critical damping in an SDOF model of a structure, then it is presumably also good enough for an MDOF model of the same structure.

Of course, there may be cases where it makes sense to use different damping ratios for different modes, for instance if dissipative elements are known to be concentrated in a part of the structure that experiences large deformations only in some modes of vibration. But we will limit ourselves to uniform modal damping in this text.

3.5.2.2 Rayleigh damping

We know that both the mass and stiffness matrices can be diagonalised by the mode shapes. Rayleigh realised that a good way to ensure that the same is true of the damping matrix is to make it a linear combination of the mass and stiffness matrices:

$$\mathbf{c} = a_0\mathbf{m} + a_1\mathbf{k} \tag{3.42}$$

Using this formulation requires choosing the coefficients a_0 and a_1 to give a suitable overall level of damping. This is most easily done by relating the Rayleigh damping to an equivalent modal damping. Following through the transformation from physical to modal co-ordinates, the generalised damping coefficient in mode i can be written as

$$\frac{C_i}{M_i} = a_0 + a_1\omega_i^2 = 2\xi_i\omega_i$$

Hence,

$$\xi_i = \frac{1}{2}\left(\frac{a_0}{\omega_i} + a_1\omega_i\right) \tag{3.43}$$

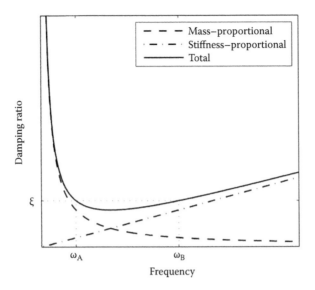

Figure 3.16 Rayleigh damping – typical variation of equivalent damping ratio with frequency.

The typical relationship between damping ratio and frequency is plotted in Figure 3.16. It can be seen that it comprises a mass-proportional term, controlled by a_0, which provides high damping at low frequencies and decays away rapidly, and a stiffness-proportional term, controlled by a_1, which increases linearly with frequency.

To choose appropriate coefficients, we can specify a desired damping level at two frequencies, say ω_A and ω_B. Substituting into Equation 3.43 will give two simultaneous equations that can be solved for a_0 and a_1. Normally we would choose the lowest and highest frequencies of interest for the system being modelled; we can specify the same damping ratio at each frequency (as has been done in Figure 3.16) or we could select different values. The graph then shows that the overall damping will be slightly lower in between the two chosen frequencies, and higher at the ends of the range. This approach will normally give a conservative estimate of the dynamic response because it causes the modes between mode A and mode B to be more lightly damped, and therefore to respond more, than modes A and B.

Rayleigh damping thus gives slightly less control than modal damping because it only allows the damping ratio to be specified exactly for two modes, the others then following automatically. But it can be particularly useful when an explicit formulation of the damping matrix is required.

3.5.2.3 Other damping formulations

More sophisticated damping formulations that still preserve the modal orthogonality property are possible. In fact, there is an infinite number of matrices that can be formed from the mass and stiffness matrices and that can be diagonalised by the mode shapes. The general form of such orthogonal damping matrices is

$$\mathbf{c} = \mathbf{m} \sum_{j=0}^{n} a_j (\mathbf{m}^{-1}\mathbf{k})^j \tag{3.44}$$

Clearly Rayleigh damping is a special case of this more general formulation, with $n = 1$. However, taking $n > 1$ allows the damping to be fixed at a greater number of frequencies than the two permitted by Rayleigh, while preserving the orthogonality property. For this more general case, following through the transformation from physical to modal co-ordinates gives the damping coefficient in mode i:

$$\frac{C_i}{M_i} = \sum_{j=0}^{n} a_j \omega_i^{2j} = 2\xi_i \omega_i$$

from which

$$\xi_i = \frac{1}{2\omega_i} \sum_{j=0}^{n} a_j \omega_i^{2j} \tag{3.45}$$

The damping ratio can be specified in as many modes as desired, and the necessary coefficients a_j then determined by solving the simultaneous equations defined by Equation 3.45. For example, setting $n = 3$ would enable us to set the damping at four modal frequencies:

$$\begin{bmatrix} \xi_1 \\ \xi_2 \\ \xi_3 \\ \xi_4 \end{bmatrix} = \frac{1}{2} \begin{bmatrix} \omega_1^{-1} & \omega_1 & \omega_1^3 & \omega_1^5 \\ \omega_2^{-1} & \omega_2 & \omega_2^3 & \omega_2^5 \\ \omega_3^{-1} & \omega_3 & \omega_3^3 & \omega_3^5 \\ \omega_4^{-1} & \omega_4 & \omega_4^3 & \omega_4^5 \end{bmatrix} \begin{bmatrix} a_0 \\ a_1 \\ a_2 \\ a_3 \end{bmatrix}$$

To find the necessary values of the coefficients a_j then requires inversion of the matrix of modal frequency terms. In principle, this method offers far more control over the damping formulation. In practice, however, Rayleigh damping is often preferred because of its simplicity, and because its accuracy is considered appropriate to the degree of certainty with which the true damping is known.

3.5.3 Implementing the mode superposition method

While implementation of the mode superposition method is generally done computationally, and is built into most finite element programs, it is as well to have an understanding of the procedure. This is set out below, assuming that modal damping is used. Each step is illustrated using an example of a simple, 3DOF shear-type building.

Step 1: Create your model of the system – decide how to represent the structure by a finite number of DOFs, formulate the mass and stiffness matrices, **m** and **k**, and the vector of time-varying nodal loads, **f**.

The model of the 3DOF shear-type building is shown in Figure 3.17, where $m = 1000$ kg, $k = 2.5$ MN/m and modal damping of 5% is assumed in all modes. The loading in this example comprises the three identical forces $F(t)$, which are periodic,

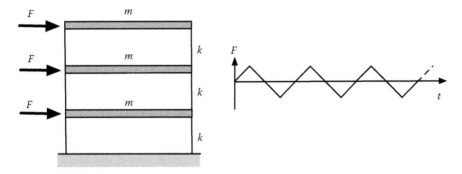

Figure 3.17 Three-storey building model and loading.

triangular-wave forces of amplitude 10 kN with frequency 3 Hz. Using Table C.1 in Appendix C, these can be approximated by the first three terms of a Fourier series as

$$F(t) = 10^4(0.811 \sin \omega t - 0.090 \sin 3\omega t + 0.032 \sin 5\omega t) \ (N)$$

where $\omega = 2\pi \times 3 = 6\pi$ rad/s.

Step 2: Find the natural frequencies ω_i and mode shapes $\mathbf{u}_i$ of the system. Usually this will be done by a numerical method such as subspace iteration, which is far more computationally efficient than explicitly solving Equation 3.12. Write the mode shape vectors as the columns of a single matrix $\mathbf{U}$.

For the structure defined in Step 1, the natural frequencies and (mass-normalised) mode shape matrix are

$$f_i = \begin{bmatrix} 3.54 \\ 9.92 \\ 14.34 \end{bmatrix} \text{Hz}, \quad \text{or} \quad \omega_i = \begin{bmatrix} 22.3 \\ 62.3 \\ 90.1 \end{bmatrix} \text{rad/s},$$

$$\mathbf{U} = \begin{bmatrix} 0.0104 & 0.0233 & -0.0187 \\ 0.0187 & 0.0104 & 0.0233 \\ 0.0233 & -0.0187 & -0.0104 \end{bmatrix}$$

As luck would have it, each natural frequency is reasonably close to the frequency of one of the harmonic terms in the loading function (which are at 3, 9 and 15 Hz).

Step 3: Formulate the generalised load vector $\mathbf{F}$, Equation 3.36. With the mode shapes mass normalised, the generalised mass matrix $\mathbf{M}$ is a unit matrix.

$$\mathbf{F} = \begin{bmatrix} 0.0104 & 0.0187 & 0.0233 \\ 0.0233 & 0.0104 & -0.0187 \\ -0.0187 & 0.0233 & -0.0104 \end{bmatrix} \begin{bmatrix} 1 \\ 1 \\ 1 \end{bmatrix} F(t) = \begin{bmatrix} 0.0524 \\ 0.0150 \\ -0.0058 \end{bmatrix} F(t)$$

where $F(t)$ is as defined in Step 1. Note that the more complicated shapes of the higher modes tend to lead to significantly smaller modal forces.

Step 4: Choose modal damping ratios ξ_i. Each mode is then governed by an equation of motion of the form of Equation 3.41. Solve these by an appropriate method from Chapter 2 (depending on the form of the loading) to give the modal displacements X_i.

In our case, the appropriate method is the procedure for calculating the response to a periodic load set out in Section 2.4.4 (Chapter 2). Applying this to each mode in turn gives the modal displacement time-histories shown in Figure 3.18. Clearly the modal displacement in the first mode is much higher than in modes 2 and 3. This is both because the loading function is dominated by a frequency component close to the first modal frequency (see Step 1) and because the conversion to modal forces further emphasises the first mode (see Step 3).

From these results, it looks as though the third mode could be omitted without much loss of accuracy, but be careful here – while this may be reasonable when looking at displacements, it is often not so when considering forces (see Step 6). Also, note that the amplitude of the first-mode response looks alarmingly large (of the order of 5 m). However, this amplitude is dependent on how we have chosen to scale the mode shapes. It is only when we multiply by the mode shape vector in the next step that we get back to physically meaningful values.

Step 5: Convert back to displacements in the physical co-ordinate system, **x**, by super-position of the modal displacements (Equation 3.17).

The resulting storey displacements are shown in Figure 3.19. As would be expected from Step 4, the displacements are dominated by the response to the first harmonic

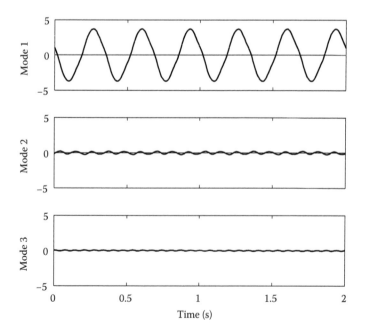

Figure 3.18 Modal displacements (in metres) of 3-storey building subjected to triangular loads.

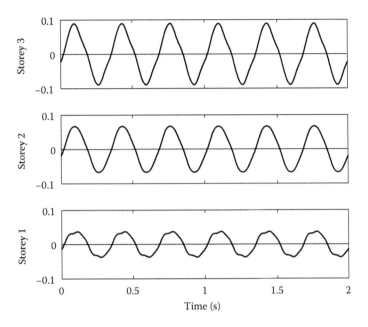

Figure 3.19 Storey displacements (in m) of 3-storey building subjected to triangular loads.

component of the loading, at 3 Hz, which is in near-resonance with the fundamental mode of the structure (3.54 Hz). However, some higher-mode contributions are visible, particularly at the first storey.

Step 6: Calculate the elastic forces $\mathbf{f}_k$ which must be resisted by the members. The elastic forces are related to the displacements via the stiffness matrix:

$$\mathbf{f}_k = \mathbf{kx} = \mathbf{kUX} = \mathbf{k}(\mathbf{u}_1 X_1 + \mathbf{u}_2 X_2 \ldots + \mathbf{u}_i X_i \ldots) \tag{3.46}$$

But from the free vibration analysis (Equation 3.13), we know that

$$\mathbf{ku}_i = \omega_i^2 \mathbf{mu}_i$$

So the elastic forces can also be written as

$$\mathbf{f}_k = \mathbf{m}\left(\omega_1^2 \mathbf{u}_1 X_1 + \omega_2^2 \mathbf{u}_2 X_2 \ldots + \omega_i^2 \mathbf{u}_i X_i \ldots\right) = \mathbf{mU.\Lambda X} \tag{3.47}$$

The matrix/vector notation can be confusing. What Equation 3.47 means is that the elastic force at DOF j due to the response of the structure in its ith mode of vibration is

$$\mathbf{f}_k(i, j) = m_j u_{ij} \omega_i^2 X_i \tag{3.48}$$

where u_{ij} is the jth term of the ith mode shape vector. To find the total force at DOF j, we sum the terms given by Equation 3.48 over all the modes of vibration i.

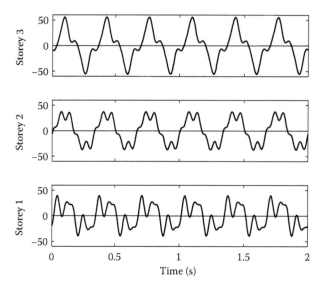

Figure 3.20 Storey forces (in kN) in 3-storey building subjected to triangular loads.

The resulting storey forces are shown in Figure 3.20. Note that multiplying the modal displacements by the frequency squared has the effect of greatly magnifying the contributions from the higher modes. As a result, the force time-histories are much more complex than the displacement ones, with the higher modes making a larger contribution. Recalling that the input force at each storey level had a magnitude of 10 kN, it can be seen that the dynamic response has resulted in storey inertia forces between 4 and 6 times larger. The multi-modal nature of the response also means that the force time-histories at each storey look quite different to each other, and that the peak forces at each storey do not occur simultaneously.

3.5.4 Mode superposition for MDOF systems with base motions

Applying the same modal transformation process to the governing equation for a MDOF system subjected to a base motion (Equation 3.8), gives

$$\mathbf{M}\ddot{\mathbf{Y}} + \mathbf{C}\dot{\mathbf{Y}} + \mathbf{K}\mathbf{Y} = -\mathbf{U}^\mathrm{T}\mathbf{m}\,\mathbf{i}\,\ddot{x}_\mathrm{g} \tag{3.49}$$

where $\mathbf{Y}$ is a vector of modal displacements satisfying $\mathbf{y} = \mathbf{U}\mathbf{Y}$ and the generalised matrices $\mathbf{M}$, $\mathbf{C}$ and $\mathbf{K}$ are as defined earlier in Equation 3.36. Dividing through by the generalised mass and neglecting the negative sign, a typical line of Equation 3.49 has the form

$$\ddot{Y}_i + 2\xi_i\omega_i\dot{Y}_i + \omega_i^2 Y_i = \frac{L_i}{M_i}\ddot{x}_\mathrm{g} \tag{3.50}$$

where L_i is called the *excitation factor* and M_i the generalised mass:

$$L_i = \mathbf{u}_i^\mathrm{T}\mathbf{m}\mathbf{i}, \quad M_i = \mathbf{u}_i^\mathrm{T}\mathbf{m}\mathbf{u}_i \tag{3.51}$$

Of course, if the mode shapes are mass normalised, then $M_i = 1$ for all i. But whatever normalisation is used, the ratio L_i/M_i is dimensionless, and is often referred to as the *modal participation factor.* Comparing Equation 3.50 to Equation 2.24 (Chapter 2), the only difference is that the input excitation has been scaled by this factor. Since the system is linear, the resulting displacement, velocity and acceleration will be scaled accordingly. So the modal participation factor can be thought of as the ratio between the response of a mode of an MDOF system and the response of an SDOF system having the same natural frequency and damping ratio.

Suppose the displacement response of an SDOF system having circular natural frequency ω_i and damping ratio ξ_i is $z_i(t)$. The corresponding modal displacement for a mode having the same natural frequency is

$$Y_i = \frac{L_i}{M_i} z_i \tag{3.52}$$

Adapting Equation 3.47, the elastic force in mode i is

$$\mathbf{f}_k(i) = \mathbf{m}\mathbf{u}_i \omega_i^2 Y_i = \mathbf{m}\mathbf{u}_i \omega_i^2 \frac{L_i}{M_i} z_i \tag{3.53}$$

So, the force at DOF j due to the response of the structure in mode i is

$$\mathbf{f}_k(i,j) = m_j u_{ij} \frac{L_i}{M_i} \omega_i^2 z_i \tag{3.54}$$

A particularly useful response parameter in earthquake analysis of buildings is the *base shear*. This is the total horizontal force that must be resisted at the foundation, and is found by summing the inertia forces over the structure. The contribution of an individual mode to the base shear is found by summing Equation 3.54 over all DOFs j, giving

$$F_b(i) = \sum_j \mathbf{f}_k(i,j) = \frac{L_i^2}{M_i} \omega_i^2 z_i \tag{3.55}$$

The ratio L_i^2/M_i is called the *effective modal mass* or the *participating mass*. It has units of mass and its sum over all modes equals the total mass of the structure. It can be thought of as the amount of mass in the building that responds in a particular mode of vibration. Both the modal participation factor and the participating mass are functions only of the mass distribution and the mode shapes – they are independent of the applied loading.

Table 3.1 shows the modal participation factors and the participating masses for the example 3DOF building introduced in Figure 3.17. Clearly the increased complexity of the higher mode shapes leads to much smaller values. Note, however, that the forces in the structure also depend on the frequency squared, which will tend to counter-act the effect of the reduced factors – see the last column of the table.

Table 3.1 Participation factors and participating masses for 3DOF building in Figure 3.17

Mode	Participation factor (L_i/M_i)	Participating mass (L_i^2/M_i)		$\omega_i^2(L_i^2/M_i)$ (10^6 kg/s^2)
		(kg)	(% of total)	
1	52.4	2742	91.4	1.36
2	15.0	225	7.5	0.87
3	−5.8	33	1.1	0.27

3.5.5 Truncation of modes

Besides decoupling the equations of motion, the mode superposition approach offers another major advantage. Often, it is possible to capture the dynamic response sufficiently accurately by summing the contributions of relatively few modes of vibration, truncating the higher modes. There are two principal reasons for this. First, since dynamic response is dominated by resonance, modes of vibration whose natural frequency is remote from the frequency range of the input loads will not be excited. Second, the more complex shapes associated with the higher modes have the effect of reducing their participation in the response to a dynamic loading (as we saw through the modal force transformation in Section 3.5.3 and the modal participation factor L_i/M_i in Section 3.5.4).

The value of this can be very significant. Consider, for example, a detailed finite element model of a structure with, say, 2000 DOFs. The full modal superposition solution would entail calculating 2000 natural frequencies and mode shapes (one for each DOF), calculating 2000 modal responses to the loading, and then converting the results back to the geometric co-ordinate system. Even with the recent huge advances in computer power, this remains a massive task. In most instances, however, it would be possible to use only a few tens of mode shapes. Whereas determining all the eigenvalues and vectors of a large matrix is highly labour-intensive, very efficient algorithms exist for extracting only the lowest modes. The resulting response calculations are also greatly simplified.

An obvious question then arises: How many modes are enough? Table 3.1 showed that, while the participation factors drop off quite quickly, the influence of the frequency multiplier on the forces tends to counter-act this, so that we risk underestimating the force demands on the structure if we truncate too soon. A good practice is to truncate the modes according to a combination of two criteria:

- Include sufficient modes to cover all significant dynamic behaviour by specifying a cut-off frequency that is some multiple of the highest significant frequency component of the loading (say two or three times), and
- Ensure that the participating mass represented by the modes is a sufficiently high fraction of the total mass (say 90 or 95%).

Example 3.11

To illustrate the mode superposition procedure for an earthquake load, and the issue of modal truncation, consider the example of the 10-storey shear building in Figure 3.21, for which the stiffness $k = 20$ MN/m and the mass $m = 10^4$ kg. The building is broadly uniform; it has uniform storey masses, except for the lighter roof, and has a modest stiffness reduction at mid-height. Take damping to be 5% of critical in all modes. Use modal superposition to determine the response of the building to the El Centro earthquake (Figure 1.20, Chapter 1).

The full set of calculations required is extensive, and must be performed by computer. We will present only edited highlights here, to illustrate the key points.

First, set up the mass and stiffness matrices and determine the undamped, free vibration properties and participation factors. The natural periods/frequencies, participation factors and participating masses are shown in Table 3.2 (based on mass-normalised mode shapes). The corresponding mode shapes are plotted in Figure 3.22, all to the same scale. For simplicity, the building has been represented as a simple linear structure with the storeys indicated by circles.

It is possible to see several *nodes* in the mode shapes – points where the modal displacement is zero; for instance, the seventh storey in mode 2. As the structure vibrates in this mode, the seventh storey remains stationary, while the structure above and below

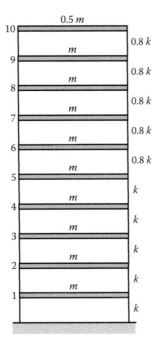

Figure 3.21 Ten-storey shear-type building.

Table 3.2 Modal properties for the 10 DOF building in Figure 3.21

Mode	Period (s)	Frequency (Hz)	Participation factor (L_i/M_i)	Participating mass (L_i^2/M_i)	
				(kg)	(% of total)
1	0.916	1.09	281.6	79,325	83.50
2	0.323	3.10	97.9	9590	10.09
3	0.193	5.18	54.5	2975	3.13
4	0.143	6.99	39.1	1526	1.61
5	0.114	8.75	−26.1	683	0.72
6	0.098	10.18	−21.7	473	0.50
7	0.088	11.40	−13.1	172	0.18
8	0.081	12.39	−11.3	127	0.13
9	0.079	12.72	8.7	75	0.08
10	0.073	13.78	7.3	54	0.06

oscillates to and fro. As we previously observed, the number of nodes in each mode shape is one less than the mode number.

For each mode, the SDOF displacement response to the earthquake loading can be calculated using Duhamel's integral, as described in Section 2.4.6 (Chapter 2), and the modal response amplitude is found by scaling the Duhamel integral output by L_i/M_i, Equation 3.52. The storey displacements are then recovered by multiplying each modal amplitude by the corresponding mode shape and summing over the modes.

While any or all storeys could have been picked, we have chosen to illustrate the response using storeys 4 and 9. Storey 4 is near the mid-height of the structure and not close to a node in any of the first few modes, so its responses might be expected to show contributions from several modes. Storey 9 is near the top and so is likely to show a

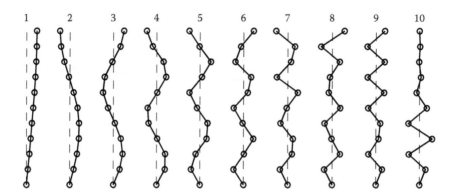

Figure 3.22 Mode shapes of 10-storey building.

relatively large response – storey 10 will have a larger peak displacement but will probably sustain smaller forces due to its reduced mass.

The displacements at storeys 4 and 9 are shown in Figure 3.23. In each case the first
15 s of the response is shown, together with a zoom of one of the peaks at around 5 s.
The solid line shows the result achieved by summing all ten modal responses, while the
lighter, dashed line (only clearly visible in the zoomed plots) shows only the first mode
response. Clearly the first mode accounts for nearly all of the displacement and little
would have been lost by omitting modes 2–10 from the calculations.

Storey forces are calculated using Equation 3.54 and are plotted in Figure 3.24 in a
similar way to the displacements. In this case, each plot contains three lines showing the
total response, the response in the first mode only, and the sum of the first two modal
responses. Significant higher frequency components are clearly visible in the force time
histories, particularly in the zoomed plots, and we can see that omitting them would lead
to a substantial underestimate of the peak force demand.

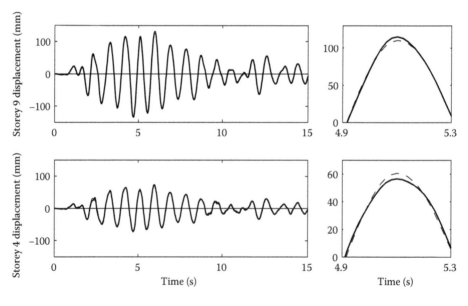

Figure 3.23 Displacements at fourth- and ninth-storey levels of 10-storey building under earthquake loading, showing first 15 s, and a zoom on the peak at around 5 s. The solid line is the full, multi-
modal response, the dashed line is calculated using the first mode only.

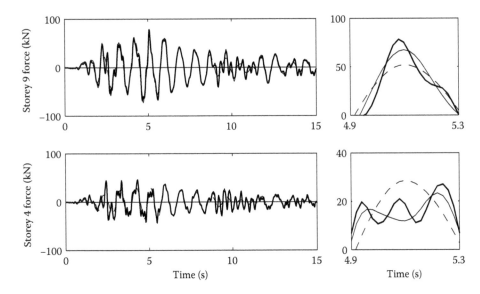

Figure 3.24 Forces at fourth- and ninth-storey levels of 10-storey building under earthquake loading, showing first 15 s and a zoom on the peak at around 5 s. The heavy, solid line is the full, multi-modal response; the dashed line is calculated using the first mode only, the lighter solid line the first two modes.

Similarly, the base shear can be calculated using Equation 3.55. This is plotted in Figure 3.25, which once again shows the result achieved using the first one, the first two and all ten modes. Here, the contributions of the higher modes are not as significant as for the storey forces. The higher modes involve different storeys accelerating (and therefore generating inertia forces) in opposite directions; they therefore tend to cancel when summed over all storeys to give the base shear.

Table 3.3 presents the modal contributions more fully, showing the accuracy achieved by adding in the contribution of each additional mode. These results have been calculated at $t = 5.96$ s, the instant of maximum displacement and base shear. Slightly different figures would be found at other times during the earthquake, but it is reasonable to focus on the instant that gives the worst values. Table 3.3 shows that including the first mode only gives a displacement estimate that is within 1% of the full 10-mode value. For the forces, three or four modes are needed to achieve this level of accuracy.

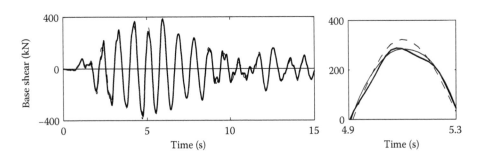

Figure 3.25 Base shear in 10-storey building under earthquake loading, showing first 15 s, and a zoom on the peak at around 5 s. The heavy, solid line is the full, multi-modal response; the dashed line is calculated using the first mode only, the lighter solid line the first two modes.

Table 3.3 Contributions of the modes to the fourth and ninth storey displacements and forces, and base shear, at instant of peak response (*t* = 5.96 s)

No. of modes included	Percentage of full modal superposition solution				
	y(4)	y(9)	F(4)	F(9)	F_b
1	**99.08**	**100.84**	**93.60**	**112.73**	**100.59**
2	100.08	100.31	**101.23**	107.99	102.25
3	99.96	99.99	**98.56**	**99.97**	100.24
4	99.99	100	**99.99**	100.45	100.09
5	100	100	100.50	100.42	100.05
6	100	100	100.38	100.18	99.99
7	100	100	100.40	100.16	99.99
8	100	100	100.34	100.10	99.99
9	100	100	100.41	99.99	99.99
10	100	100	100	100	100

Note: An error of less than 1% is achieved by including only the modal contributions indicated in bold.

The final comparison we can do is to plot profiles of displacement and force over the height of the structure at any instant in time, to check that the time history results we have plotted for storeys 4 and 9 are representative. This is done at *t* = 5.96 s in Figure 3.26. This tells a similar story to the earlier results; while the displacement profile as estimated by the first mode is virtually indistinguishable from that calculated using all 10 modes, the force profiles differ markedly. The multi-modal force distribution shows significant influence of the higher modes, resulting in higher forces in storeys 5–7 and slightly lower ones near the top and bottom of the building.

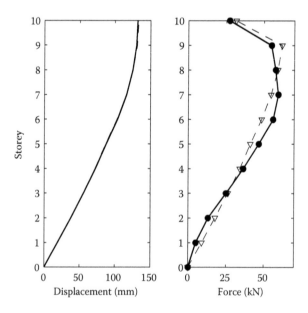

Figure 3.26 Profiles of displacement and storey forces at *t* = 5.96: first mode only (dashed/triangles) and full solution (solid/circles).

3.5.6 Mode superposition and response spectrum analysis

We saw in Chapter 2 that a response spectrum offers an attractive alternative to a full dynamic analysis under a specific earthquake time history. A response spectrum analysis takes us directly to the maximum structural response without the need to compute variations throughout the earthquake, and a smoothed design spectrum neatly summarises the likely peak response to an ensemble of earthquake time histories. If we could apply the same approach to the analysis of MDOF systems, it would greatly reduce the required computations compared to analyses such as that in Example 3.11.

The use of response spectra in mode superposition analysis is a widespread practice in the industry. It generally yields acceptable results, although it has one important limitation – because the response spectrum analysis only gives us the peak response parameters in each mode and tells us nothing about when they occur, it becomes difficult to combine them. We have seen in previous examples that the peak responses in different modes of vibration are unlikely to occur at the same time. Therefore, simply adding the peak modal responses is likely to result in a significant overestimate of the total response. Instead, an approximate combination method must be used to give an estimate of the likely peak overall response. Various such methods have been proposed of which the most widely accepted is the square root of sum of squares (SRSS) method. Take, for example, the displacement y_j at DOF j of a structure. If our analysis has yielded peak responses $y_{ji}(\text{max})$ in each mode i, then the SRSS estimate of the maximum total displacement is

$$y_j(\text{max}) = \sqrt{y_{j1}(\text{max})^2 + y_{j2}(\text{max})^2 + y_{j3}(\text{max})^2 \ldots} = \sqrt{\sum_{i=1}^{n} y_{ji}(\text{max})^2} \tag{3.56}$$

Thus, the only significant changes of approach we need to make are that we determine peak modal responses using the response spectrum rather than Duhamel's integral, and we combine peak modal responses using Equation 3.56 rather than summing the modal responses at each point in time. In other respects, the process is the same as in Section 3.5.4. Obviously, the modal combination method is approximate, but the degree of error introduced is considered acceptable in the context of the numerous other elements of uncertainty in the analysis. To illustrate the response spectrum approach, we will revisit some elements of the calculations presented in Example 3.11.

Example 3.12

Consider the structure of Example 3.11, subjected to an earthquake defined by the Eurocode 8 Type I response spectrum for ground type C plotted in Figure 2.22b, scaled to a peak ground acceleration of 3.1 m/s². Estimate the peak displacements of floors 4 and 9 and the maximum base shear.

Natural periods and participation factors are unchanged from the previous analysis, and are listed in Table 3.2. The further calculations needed here are summarised in Table 3.4. First, spectral accelerations corresponding to each natural period can be read from the response spectrum. The values obtained from Figure 2.22b (Chapter 2) need to be scaled by the PGA of 3.1 m/s². Remember that this is the peak acceleration experience by an SDOF system having the same natural period as the mode in question.

Table 3.4 Response spectrum calculation for the 10 DOF building in Figure 3.21

Mode i	Period (s)	$S_a(i)$ (m/s²)	$F_b(i) =$ $(L_i^2/M_i) \cdot S_a(i)$ (kN)	$Y_{i,max}$ (mm)	$y_{4,i}(max)$ (mm)	$y_{9,i}(max)$ (mm)
1	0.916	5.8	463.1	34.94	87.3	158.6
2	0.323	8.9	85.5	2.31	9.9	−9.3
3	0.193	8.7	26.0	0.45	0.3	1.4
4	0.143	7.4	11.3	0.15	−0.5	−0.3
5	0.114	6.6	4.5	−0.06	−0.2	0
6	0.098	6.2	2.9	−0.03	0	0
7	0.088	5.9	1.0	−0.02	0	0
8	0.081	5.7	0.7	−0.01	0	0.1
9	0.079	5.7	0.4	0.01	0	0
10	0.073	5.5	0.3	0.01	0	0
SRSS			471.8		87.9	158.9

Next we can calculate the base shear in each mode. In Equation 3.55, the base shear was given in terms of an SDOF displacement $z_i(t)$. Noting that

$$\omega_i^2 z_i(t) = \ddot{z}_i(t)$$

we can recast Equation 3.55 as

$$F_b(i)_{max} = \frac{L_i^2}{M_i}\left(\omega_i^2 z_i\right)_{max} = \frac{L_i^2}{M_i}S_a(i)$$

We next use Equation 3.52 to find the displacement amplitude of each mode:

$$Y_{i,max} = \frac{L_i}{M_i}z_{i,max} = \frac{L_i}{M_i}\frac{S_a(i)}{\omega_i^2}$$

The maximum modal displacements at floors 4 and 9 are obtained by multiplying $Y_{i,max}$ by the fourth and ninth elements of the vector for mode shape i. Finally, the overall results are obtained by SRSS combination of the values for each mode. It can be seen that the displacements are completely dominated by the response in the first mode, while for the base shear inclusion of some higher modes makes a modest difference to the result.

Comparing with the peak results obtained from the time domain analysis of Example 3.11, we see that the response spectrum analysis gives somewhat higher values: a base shear of 472 kN compared with 383 kN, a fourth-storey displacement of 88 mm compared with 73 mm, and a ninth-storey displacement of 159 mm compared with 131 mm. This is largely due to the fact that the response spectrum used in this example represents a somewhat more onerous loading than the El Centro time history. Given the relatively small values of the higher mode response parameters, it is unlikely that a large part of the discrepancy is due to the modal combination method – though it may be a source of inaccuracy in an analysis that yielded larger higher mode responses.

KEY POINTS

- As with SDOF systems, dynamic analysis of an MDOF system should normally commence with a free vibration analysis before proceeding to the analysis of response to a dynamic load.
- A linear n-DOF system is capable of vibrating in n modes, each having a characteristic deformation shape associated with a particular vibration frequency.
- The modes of vibration have useful mathematical properties that can be used to diagonalise the mass and stiffness matrices.
- Damping in MDOF systems may be conveniently modelled by assigning a damping ratio to each mode of vibration, or using the Rayleigh approach of assigning mass- and stiffness-proportional damping terms.
- The most useful, versatile analysis method for linear MDOF systems is mode superposition, in which the total response is written as the sum of the modal responses, which can be found using SDOF methods.
- When using modal superposition, it is often found that the response is dominated by only a few of the modes, and the remainder can be omitted (truncated).
- Analysis of lumped MDOF systems involves the solution of large matrix equations and is usually impractical to do by hand.
- For seismic analysis of MDOF structures, it is possible to use the response spectrum approach introduced in Chapter 2, but an approximate method must be adopted to combine the peak modal responses, which do not all occur at the same instance in time.
- In some instances, resonant response of a structure can be greatly reduced by the use of a tuned mass damper (TMD) – a secondary system tuned to the same natural frequency as the main structure.

TUTORIAL PROBLEMS

1. For the mass-spring system in Figure P3.1, (a) write down the equations of motion for the two masses in matrix form, (b) find the natural frequencies and (c) find the corresponding mode shapes.

2. Show that the mass-spring system in Figure P3.2 has the same natural frequencies as that in Figure P3.1.

3. Figure P3.3 shows two pendulums of mass m and length L, connected by a spring of stiffness k. Assuming small angles of rotation (such that $\cos \theta \approx 1$ and $\sin \theta \approx \theta$), find the natural frequencies and the corresponding mode shapes. If $m = 1$ kg and $L = 1$ m, choose the value of k such that the ratio of the second natural frequency to the first is (a) 2, (b) 10.

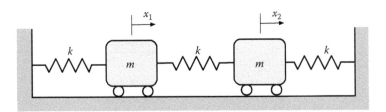

Figure P3.1

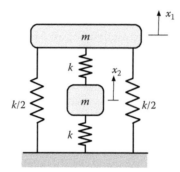

Figure P3.2

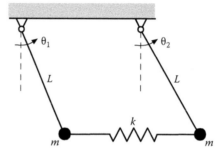

Figure P3.3

4. Suppose that, in the system of Figure P3.1, the middle spring is replaced by a damper with coefficient c. Derive the new equations of motion and comment on the effect of this change on the determination of natural frequencies and mode shapes.

5. Figure P3.5 shows an idealisation of a 3-storey building subject to sway vibrations. Write down the equations of motion and the characteristic equation from which the natural frequencies and mode shapes can be determined.

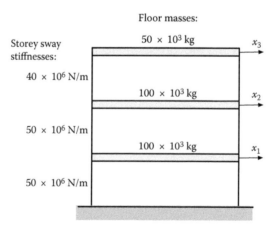

Figure P3.5

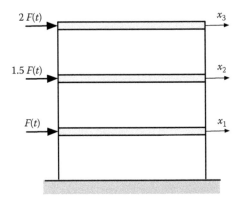

Figure P3.8

6. The natural frequencies of the structure in Figure P3.5 are $\omega_1 = 11.51$ rad/s, $\omega_2 = 30.20$ rad/s, $\omega_3 = 40.69$ rad/s. Find the corresponding mode shapes.

7. For the structure in Figure P3.5, choose Rayleigh damping parameters a_0 and a_1 to give 5% of critical damping in the first and third modes of vibration and find the resulting damping ratio for the second mode.

8. Figure P3.8 shows the same structure as in Problems 5–7, but now loaded by a set of forces $F(t)$ where

$F(t) = F_0 \sin (2\pi f t), \quad F_0 = 200 \text{ kN}, \quad f = 2 \text{ Hz}$

Find the generalised forces for the three modes calculated in Problem 6. Using the harmonic load solution from Section 2.4.2, find the peak storey displacements in the first mode, assuming 5% damping.

Chapter 4

Continuous systems

4.1 INTRODUCTION

While most practical dynamic analysis involves lumped-parameters systems, whether in reality or because of discretisation by a finite element program, it is also possible to perform dynamic analysis of structures or elements using the methods of continuum mechanics. Today's engineer is unlikely to spend much time doing this, so the presentation of the topic here will be kept relatively brief, and restricted to beams. What can be very useful, however, is to be able to reduce a continuum structure to a simple, representative, lumped-parameter equivalent that lends itself to the analysis methods of the earlier chapters. This aspect will therefore be pursued in a little more depth.

This chapter follows the by-now familiar pattern of first establishing the governing equation of motion of the system, then solving the undamped free vibration case, and only then moving on to consider the dynamic response to a particular forcing function. At the end, you should be able to

- Use the methods of continuum mechanics to derive the governing equations for beam vibrations,
- Find natural frequencies and mode shapes for uniform beams by applying appropriate boundary conditions to the general solution,
- Analyse the vibration response of beams to simple forms of dynamic loading,
- Use an approximation of the vibration shape of a continuous element to derive an equivalent single-degree-of-freedom (SDOF) model,
- Understand the limitations and possible sources of inaccuracy of the equivalent SDOF approach.

4.2 VIBRATIONS OF BEAMS

4.2.1 Undamped equation of motion

Consider an initially straight, horizontal beam subjected to vertical loading and undergoing vibrations in the vertical plane (Figure 4.1). We define the position along the beam by the co-ordinate z. To keep the analysis general at this stage, we allow the flexural rigidity $EI(z)$ and the mass per unit length $m(z)$ to vary along the beam, and the loading per unit length $p(z, t)$ to vary with both position and time. The support conditions need not be specified yet – they are shown as simple supports for illustrative purposes only.

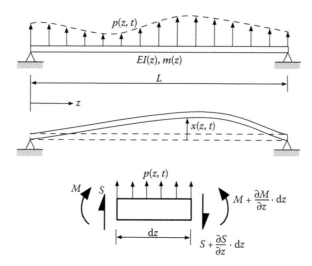

Figure 4.1 Flexural vibrations of a beam: general layout and free-body diagram for a short element.

As with lumped-parameter systems, the first step in a dynamic analysis is to set up the equation of motion. In doing so, we will neglect three effects:

- Damping – it is possible to derive an equation of motion including damping, but it can be difficult to produce a realistic model of the distributed damping within a beam, and the governing equations become hard to solve. For our purposes, damping is better incorporated by the use of a damping ratio in the dynamic response calculation, as was done for lumped-parameter systems in Chapters 2 and 3.
- Rotary inertia effects – unless it lies on a vertical axis of symmetry, as a beam cross-section vibrates vertically it also rotates slightly, resulting in rotary inertia forces equal to the mass moment of inertia of the beam multiplied by the rotational acceleration.
- Shear deformations – shear forces in the beam cause distortions, which contribute to the vertical deflection, but are usually much smaller than flexural deflections. Shear deformations also influence the amount of rotation of a cross-section, and so interact with rotary inertia effects in a rather complex way.

These last two effects can be significant in relatively short, stocky beams but it is reasonable to neglect them in most instances.

Using the normal continuum mechanics approach, we isolate a very short section of beam of length dz and consider the forces acting on it, as shown in Figure 4.1. Resolving vertically:

$$p.dz - \frac{\partial S}{\partial z}dz = (m.dz)\frac{\partial^2 x}{\partial t^2} \quad \rightarrow \quad p - \frac{\partial S}{\partial z} = m\frac{\partial^2 x}{\partial t^2} \tag{4.1}$$

where the curly ∂ denotes a partial differential, that is, differentiation of a function of more than one independent variable (in this case z and t) with respect to one of those variables. Taking moments about an axis normal to the page at the right-hand end and noting that, as dz is small, $dz^2 \ll dz$:

$$\frac{\partial M}{\partial z}dz = S.dz + p.dz\frac{dz}{2} \quad \rightarrow \quad \frac{\partial M}{\partial z} = S \tag{4.2}$$

From elementary beam theory, the moment is related to the curvature (second differential of deflected shape) by:

$$M = EI \frac{\partial^2 x}{\partial z^2} \tag{4.3}$$

Combining Equations 4.1 to 4.3 gives the equation of motion, often known as Euler's equation:

$$\frac{\partial^2}{\partial z^2}\left(EI \frac{\partial^2 x}{\partial z^2} \right) + m \frac{\partial^2 x}{\partial t^2} = p \tag{4.4}$$

In many practical cases, EI is constant and so can be taken outside the differential giving

$$EI \frac{\partial^4 x}{\partial z^4} + m \frac{\partial^2 x}{\partial t^2} = p \tag{4.5}$$

4.2.2 Undamped free vibration of uniform beams

As with other vibration problems we have encountered, the normal analysis approach is to begin by considering the undamped free vibration case, that is, with the loading set to zero. At this stage, we will also simplify the analysis by assuming that neither EI nor m vary with z, so that the governing partial differential equation has constant coefficients:

$$EI \frac{\partial^4 x}{\partial z^4} + m \frac{\partial^2 x}{\partial t^2} = 0 \tag{4.6}$$

The standard method of solution is separation of variables, in which the displacement x, which varies with both position and time, is written as the product of two functions, one in terms of position only and the other in terms of time only. The full procedure is explained in Appendix F and yields the solution:

$$x(z, t) = u(z).y(t) \tag{4.7}$$

where

$$u(z) = B_1 \sin \alpha z + B_2 \cos \alpha z + B_3 \sinh \alpha z + B_4 \cosh \alpha z \tag{4.8}$$

$$y(t) = A_1 \sin \omega_n t + A_2 \cos \omega_n t \tag{4.9}$$

$$\omega_n^2 = \frac{\alpha^4 EI}{m} \quad \text{or} \quad \alpha = \left(\frac{\omega_n^2 m}{EI} \right)^{1/4} \tag{4.10}$$

The position function $u(z)$ is the mode shape. Imposition of four boundary conditions will enable determination of the coefficients B_1 to B_4 and the parameter α, and hence ω_n. A distributed-parameter system has an infinite number of DOFs (think of it as an assemblage of an infinite number of infinitesimally small lumped masses). Since one mode of vibration can be found for each DOF, it follows that there will be an infinite sequence of modes of vibration. The application of the boundary conditions will result in an infinite series of natural frequencies ω_i, each associated with a particular mode shape u_i. As with lumped, MDOF systems, the scaling of the mode shape is arbitrary because application of the boundary conditions will always leave one of B_1 to B_4 undetermined. Also in common with MDOF systems, we generally find that the higher modes have shapes that are more complex and that the response to most loadings can be adequately captured by including only a limited number of modes.

Having determined the natural frequencies and mode shapes, the amplitude of the vibrations can be determined by imposition of the initial conditions to give the time-function coefficients A_1 and A_2.

Example 4.1

Determine the natural frequencies and mode shapes of a simply supported, uniform beam of length L, mass per unit length m and flexural rigidity EI.

A simply supported beam has pinned supports, giving zero deflection and moment at each end, hence the four boundary conditions can be written as

$$x(0) = M(0) = x(L) = M(L) = 0$$

From the moment-curvature relation, Equation 4.3, the moment conditions imply (using a dash notation to indicate differentiation with respect to z):

$$x''(0) = x''(L) = 0$$

Since the four boundary values must occur at all values of time, it follows that they must be achieved by the mode shape, independent of the time function, that is, we must have

$$u(0) = u''(0) = u(L) = u''(L) = 0$$

The general form of the mode shape is given by Equation 4.8 and its second differential is

$$u'' = \alpha^2(-B_1 \sin \alpha z - B_2 \cos \alpha z + B_3 \sinh \alpha z + B_4 \cosh \alpha z)$$

Inserting each of the four boundary conditions in turn gives:

$$B_2 + B_4 = 0$$
$$-B_2 + B_4 = 0$$
$$B_1 \sin \alpha L + B_3 \sinh \alpha L = 0$$
$$-B_1 \sin \alpha L + B_3 \sinh \alpha L = 0$$

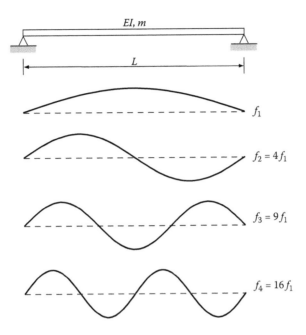

Figure 4.2 The first four modes of a simply supported beam.

This leads to $B_2 = B_3 = B_4 = 0$. The last two equations then leave us with

$B_1 \sin \alpha L = 0$

Either $B_1 = 0$, in which case the beam does not move, or B_1 cannot be determined and $\sin \alpha L = 0$, which requires

$\alpha_i L = i\pi, \quad i = 1, 2, 3...$

Since the mode shape scaling is arbitrary, the simplest thing to do is to put $B_1 = 1$ so that the mode shapes can be written as

$$u_i = \sin \frac{i\pi z}{L}$$

and the corresponding natural frequencies are found by substituting α into Equation 4.10:

$$\omega_i = \left(\frac{i\pi}{L} \right)^2 \sqrt{\frac{EI}{m}} = i^2 \pi^2 \sqrt{\frac{EI}{mL^4}} \rightarrow f_i = \frac{i^2 \pi}{2} \sqrt{\frac{EI}{mL^4}}$$

The first four modes are shown in Figure 4.2. Here it can be seen that the mode number simply defines the number of half-sine waves in the mode shape.

Other support cases can be solved in a similar way, though the algebra does get rather more complicated. Table 4.1 summarises the modal properties of uniform beams with the most common support conditions, and the first four mode shapes for the cases other than pinned-pinned are plotted in Figure 4.3. Both Figures 4.2 and 4.3 follow the pattern established for MDOF systems in Chapter 3, that the number of nodes (stationary points) in the mode shape is one less than the mode number.

Table 4.1 Modal parameters for uniform beams with various end conditions (see Equation 4.8 for definitions of α and $B_1 - B_4$)

End conditions	Mode	αL	$f \times \sqrt{\dfrac{EI}{mL^4}}$	Mode shape coefficients (arbitrary normalisation)			
				B_1	B_2	B_3	B_4
Pinned-pinned	1	3.142	1.571	1	0	0	0
	2	6.283	6.283				
	3	9.425	14.14				
	4	12.57	25.13				
Fixed-free	1	1.875	0.560	1	$-B$	-1	B
	2	4.694	3.507	where $B = \dfrac{\sin \alpha L + \sinh \alpha L}{\cos \alpha L + \cosh \alpha L}$			
	3	7.855	9.820				
	4	11.00	19.24				
Fixed-fixed	1	4.730	3.561	1	B	-1	$-B$
	2	7.853	9.815	where $B = \dfrac{\cos \alpha L - \cosh \alpha L}{\sin \alpha L + \sinh \alpha L}$			
	3	10.99	19.23				
	4	14.14	31.81				
Fixed-pinned	1	3.940	2.471	1	$-B$	-1	B
	2	7.068	7.951	where $B = \tanh \alpha L$			
	3	10.21	16.59				
	4	13.35	28.37				

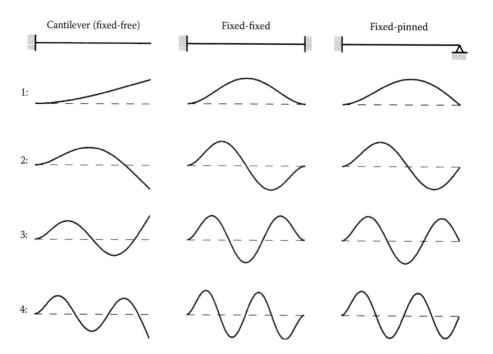

Figure 4.3 The first four mode shapes of cantilever, fixed-ended and fixed-pinned beams.

Example 4.2

Returning to the uniform, simply supported beam in Example 4.1, suppose that the application of a symmetrically distributed static load causes an initial displaced shape, which varies quadratically with position and takes a maximum value X at midspan. If the load is suddenly removed, describe the resulting free vibrations.

As the beam is released from an initial displacement, we would expect the ensuing vibrations to vary as the cosine of time (following the same rule as established in earlier chapters, that an initial displacement produces a $\cos t$ response, while an initial velocity results in a $\sin t$ response). In Example 4.1, we found an infinite series of modes for a simply supported beam and the total motion can be written as a linear combination of the vibrations in each mode:

$$x(z,t) = A_1 \sin\frac{\pi z}{L}\cos\omega_1 t + A_2 \sin\frac{2\pi z}{L}\cos\omega_2 t + A_3 \sin\frac{3\pi z}{L}\cos\omega_3 t \dots \tag{4.11}$$

It is quite easy to derive the quadratic function describing the initial shape. Taking a general quadratic of the form $x(z, 0) = az^2 + bz + c$ and putting in the co-ordinates of the known points: $x(0, 0) = 0$, $x(L/2, 0) = X$, $x'(L/2, 0) = 0$ (from symmetry) gives:

$$x(z,0) = 4X\left(\frac{z}{L} - \frac{z^2}{L^2}\right)$$

To use this initial condition to find the modal amplitude coefficients A_i, it would be helpful if the two above expressions looked more similar. If we assume the initial shape repeats outside the range $[0, L]$ in alternating positive and negative cycles so as to produce an odd periodic function (i.e. one for which $x(-z) = -x(z)$), as shown in Figure 4.4, then we can write it as a Fourier sine series. Details of how to do this are given in Appendix C. We get:

$$x(z,0) = 4X\left(\frac{8}{\pi^3}\sin\frac{\pi z}{L} + \frac{8}{27\pi^3}\sin\frac{3\pi z}{L}\dots\right) \tag{4.12}$$

Of course, each of these Fourier terms looks like one of the mode shape terms in Equation 4.11. Note that the even Fourier terms are all zero – this is to be expected because they are anti-symmetric over the range $[0, L]$ and we are approximating a function that is symmetric over that interval. Also, we see that the coefficients get small very quickly – the first sine term gives quite a good approximation of the quadratic function,

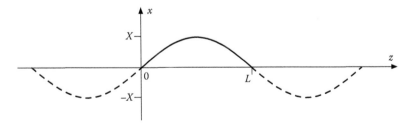

Figure 4.4 Assumed extrapolation of a beam's initial shape outside the range $0 - L$, to enable its representation by a Fourier sine series.

only a small contribution is needed from the next term to improve it, and subsequent ones are smaller still. If we now put $t = 0$ in our general solution, Equation 4.11 above, we get:

$$x(z, 0) = A_1 \sin\frac{\pi z}{L} + A_2 \sin\frac{2\pi z}{L} + A_3 \sin\frac{3\pi z}{L} \dots \tag{4.13}$$

Equating coefficients in Equations 4.12 and 4.13 gives

$$A_1 = \frac{32X}{\pi^3}, \quad A_2 = 0, \quad A_3 = \frac{32X}{27\pi^3} \dots$$

so the full solution can be written as

$$x(z, t) = \frac{32X}{\pi^3}\left(\sin\frac{\pi z}{L}\cos\omega_1 t + \frac{1}{27}\sin\frac{3\pi z}{L}\cos\omega_3 t \dots\right)$$

The full solution comprises a linear combination of the odd-numbered modes of vibration; the symmetrical initial deflected shape means that the even-numbered (anti-symmetric) modes are not excited at all. Motion is dominated by the fundamental mode, in which the beam deforms as a single half-sine wave. The third mode has a displacement amplitude of only 1/27 of the fundamental, and higher modes will respond considerably less than this.

4.2.3 Forced vibration response

The response of a continuous system to a dynamic forcing function can be determined using a mode superposition approach directly analogous to that used for lumped MDOF structures (Section 3.5, Chapter 3). We write the dynamic displacement as a linear combination of the mode shapes and some time-varying amplitude functions:

$$x(z, t) = \sum_{i=1}^{\infty} u_i(z).X_i(t) \tag{4.14}$$

By a similar process to that of Section 3.5 (Chapter 3), this results in a series of SDOF equations for the modal amplitudes, which can most conveniently be written as

$$\ddot{X}_i + \omega_i^2 X_i = \frac{P_i(t)}{M_i} \tag{4.15}$$

where the modal mass M_i and modal force P_i are given by

$$M_i = \int_0^L m(z)u_i^2(z)\,dz, \quad P_i(t) = \int_0^L p(z,t)u_i(z)\,dz \tag{4.16}$$

Here, the continuous nature of the analysis means that the modal properties are defined by integrals rather than discrete summations, as would be the case for a lumped system. The exception is when concentrated loads are applied at specific points on the beam, where the modal force is simply the product of force and the mode shape at its point of application.

If desired, damping can be included in the analysis by specifying a modal damping ratio ξ_i (again exactly as in the lumped MDOF case), so that Equation 4.15 becomes

$$\ddot{X}_i + 2\xi_i\omega_i\dot{X}_i + \omega_i^2 X_i = \frac{P_i(t)}{M_i} \tag{4.17}$$

These SDOF equations can be solved in the usual way and the solutions are then combined using Equation 4.14 to find the full solution. Usually a good approximation to the exact solution can be obtained with only the first few modes. Structural accelerations (if required) can be obtained by differentiating Equation 4.14 to give

$$\ddot{x}(z, t) = \sum_{i=1}^{\infty} u_i(z).\ddot{X}_i(t) \tag{4.18}$$

Besides calculating the motions, a structural analysis will normally also have the objective of finding the internal forces to be resisted. For beams, the forces of greatest interest are generally the bending moments. Combining Equations 4.3 and 4.14 gives

$$M(z, t) = EIx''(z, t) = EI\sum_{i=1}^{\infty} u_i''(z).X_i(t) \tag{4.19}$$

Example 4.3

A simply supported beam of span $L = 7.5$ m has a flexural rigidity $EI = 20$ MNm^2 and supports a mass per unit length $m = 1000$ kg/m. It is loaded by a concentrated, sinusoidal force of amplitude 1000 N and frequency $f_F = 5$ Hz, at 2.0 m from one support. Assuming 2% damping in all modes, determine the steady-state dynamic response of the beam.

Step 1: Modal analysis. Following the same procedure as in Example 4.1, the first five natural frequencies are $f_1 = 3.9$, $f_2 = 15.8$, $f_3 = 35.5$, $f_4 = 63.2$, $f_5 = 98.7$ Hz. In each case, the mode number equals the number of half-sine waves in the mode shape. Assume the mode shapes are normalised to a peak value of 1.

Step 2: Calculate the modal masses and forces (Equation 4.16). With the mode shapes as simple sinusoids with a peak value of 1, the modal masses are all equal to half the total mass, $M_i = 3750$ kg.

For the modal forces, with only a point force it is not necessary to perform an integral – each modal force is simply given by the product of the force and the mode shape amplitude at the point of force application (the time-varying part is the same in all cases). This gives

$$P_1 = 743, P_2 = 995, P_3 = 588, P_4 = -208, P_5 = -866 \text{ N}$$

Step 3: Solve the modal SDOF equations (Equation 4.17). With a harmonic load of frequency 5 Hz, the steady state responses in each mode can be found from Equations 2.29 to 2.31 (Chapter 2) as:

$$X_i = A_i \sin(2\pi f_F t - \phi_i)$$

where:

$A_1 = 0.532$, $A_2 = 0.030$, $A_3 = 0.0032$, $A_4 = -0.0004$, $A_5 = -0.0006$ mm

$\phi_1 = 3.06$, $\phi_2 = 0.014$, $\phi_3 = 0.006$, $\phi_4 = 0.003$, $\phi_5 = 0.002$ rads

Since we are looking only at the steady state response (the free vibration part of the response being assumed to have decayed away), the response in all modes is at the loading frequency. The differences between the modes are in the amplitudes and phases of the responses. The first mode is below the forcing frequency and has a phase angle of almost π. The other modes are all well above the forcing frequency and have relatively small amplitudes, and phase angles close to zero. The higher mode responses are therefore in phase with each other and with the loading, and almost exactly out of phase with the first mode.

Step 4: Combine the modal solutions to give the full displacement solution (Equation 4.14). This can be written as

$$x(z, t) = \sum_{i=1}^{\infty} \sin\frac{i\pi z}{L}.A_i \sin(2\pi f_F t - \phi_i)$$

where, for the first five modes, values for A_i and ϕ_i were found in Step 3. Figure 4.5 shows the displaced shape of the beam at the first positive vibration peak ($t = 0.148$ s), and the time history at the point of maximum displacement ($z = 4.01$ m).

Compared to a static analysis, the displaced shape may seem a little surprising. With the load 2 m from the left-hand end, the static deflected shape would have a peak slightly to the left of centre; however, our dynamic analysis produces the opposite result, with the peak displacement slightly to the right of centre. The cause of this can be understood by looking at the contributions of the individual modes. Here, the phase differences between the modes are significant. The first mode is well below the load frequency and so its response is almost completely out of phase with the loading – a phase angle of 3.06 rads, or about 175°. The other modes all have frequencies quite a bit higher than the loading frequency, and so respond almost exactly in phase with the load, and therefore out of phase with the mode 1 response. At $t = 0.148$ s, the modal contributions to the displaced shape

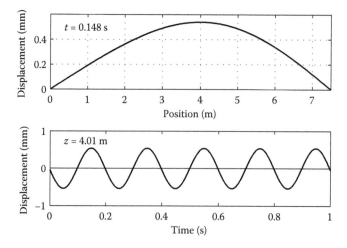

Figure 4.5 Displacement response of a simply supported beam to a harmonic point load at $z = 2$ m.

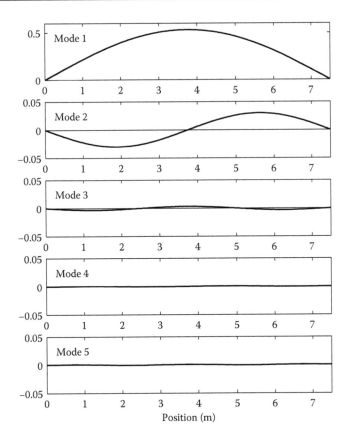

Figure 4.6 Modal contributions to displacement (mm) at $t = 0.148$ s. (Note that modes 2–5 are plotted to a larger vertical scale than mode 1.)

are as shown in Figure 4.6. Including several modes introduces only a small change in shape compared to using a single-mode approximation, shifting the location of the peak displacement to the right a little. Clearly, the difference between the first-mode contribution and the total displacement in Figure 4.5 can be explained principally by the contribution from the second mode.

Step 5: Compute the bending moments according to Equation 4.19:

$$M(z, t) = -\frac{EI\pi^2}{L^2} \sum_{i=1}^{\infty} i^2 \sin\frac{i\pi z}{L}.A_i \sin(2\pi f_F t - \phi_i)$$

Notice that differentiating the mode shape results in the appearance of the i^2 factor, which will cause the higher modes to have a greater influence on the bending moments than on the displacements. Figure 4.7 shows the bending moment diagram at the instant of the maximum positive moment ($t = 0.248$ s), and the time history at the point of maximum moment ($z = 4.28$ m). The bending moment plot also shows the moment distribution due to a static load at $z = 2$ m. As expected, the influence of the higher modes is more pronounced than for the displacements. Similar to the displacements, the phase differences between the higher modes and the fundamental result in a reduction in the amplitude of the response around the load point, producing a moment distribution that is markedly different from that due to a static load.

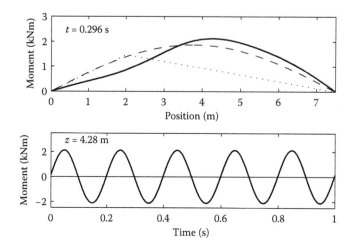

Figure 4.7 Dynamic bending moments in a simply supported beam due to a harmonic point load at $z = 2$ m. (Solid lines: sum of first five modal responses; dashed line: first mode only, dotted line: bending moment diagram due to static load at $z = 2$ m.)

The results of this analysis are, of course, highly sensitive to the loading frequency. In the example chosen, this is not particularly close to any of the modal frequencies and so there are no large dynamic amplifications. Also, because the load frequency is higher than that of one mode but much lower than the others, the first and higher modal responses are almost in anti-phase, which results in the somewhat counter-intuitive distribution of displacements and moments.

If the load frequency were close to that of the fundamental mode (around 4 Hz), then resonance in that mode would dominate the response, giving near-symmetrical distributions of displacement and moment, despite the off-centre loading. Conversely, if the loading frequency was increased toward that of the second mode (around 15 Hz), the second modal response would come to dominate, resulting in larger displacements around the load point and the quarter-span points than at midspan.

4.3 EQUIVALENT SINGLE-DEGREE-OF-FREEDOM SYSTEMS

Clearly, the preceding analysis quickly becomes cumbersome. For anything other than a simply supported beam, the exact mode shapes are quite complex and the response calculations are lengthy. Often a good approximation to the behaviour can be achieved by reducing the continuous system to an equivalent SDOF system.

4.3.1 Generalised properties

Return to the general beam setup of Figure 4.1, in which the mass, flexural rigidity and load may vary with z. To approximate this as an SDOF system, we have to assume that the structure may only deflect in a single shape – very often this will be our best estimate of the fundamental mode shape, but it need not necessarily be so. For the sake of generality, we will just call it the *shape function*, denoted by $\psi(z)$. The actual displacement of the structure is obtained by scaling ψ by a time-varying *generalised displacement* $X(t)$:

$$x(z, t) = X(t).\psi(z) \tag{4.20}$$

The generalised displacement is usually chosen as the actual displacement of some convenient point on the structure, such as the midpoint of a simply supported beam or the tip of a cantilever. It follows that the shape function must be normalised to a value of unity at this point. Equation 4.20 is easily differentiated with respect to either variable:

$$\dot{x} = \dot{X}\psi, \quad \ddot{x} = \ddot{X}\psi; \quad x' = X\psi', \quad x = X\psi'' \tag{4.21}$$

where a dash denotes differentiation with respect to position z, and a dot with respect to time t.

Now use an energy approach to formulate the equation of motion. Over a short time interval dt the structure undergoes a small displacement $dx(z, t)$. The increment of work dW done by the applied load $p(z, t)$ as it moves through the distance dx is

$$dW = \int_0^L p(z,t).dx(z,t)\,dz = \int_0^L \left(p.\frac{\partial x}{\partial t} dt \right) dz = \int_0^L (p\dot{x}.dt)\,dz$$

Therefore, substituting from Equation 4.20:

$$\frac{dW}{dt} = \int_0^L p\dot{x}\,dz = \int_0^L p\dot{X}\psi\,dz \tag{4.22}$$

If we neglect damping for the time being, the energy U of the system is a combination of kinetic and flexural strain energies:

$$U = \int_0^L \frac{1}{2}m\dot{x}^2\,dz + \int_0^L \frac{1}{2}EIx''^2\,dz = \int_0^L \frac{1}{2}m\dot{X}^2\psi^2\,dz + \int_0^L \frac{1}{2}EIX^2\psi''^2\,dz \tag{4.23}$$

With no damping, and therefore no energy losses, the rate of change of energy equals the rate of work done by the load:

$$\frac{dU}{dt} = \frac{dW}{dt}$$

Substituting from Equations 4.22 and 4.23 gives

$$\int_0^L m\dot{X}\ddot{X}\psi^2\,dz + \int_0^L EIX\dot{X}\psi''^2\,dz = \int_0^L p\dot{X}\psi\,dz$$

which, on removal of the common factor $\dot{X}$, becomes

$$\int_0^L m\psi^2\ddot{X}\,dz + \int_0^L EI\psi''^2 X\,dz = \int_0^L p\psi\,dz \tag{4.24}$$

or

$$m^* \ddot{X} + k^* X = p^* \tag{4.25}$$

where the terms

$$m^* = \int_0^L m \psi^2 \, dz, \quad k^* = \int_0^L EI \psi''^2 \, dz, \quad p^* = \int_0^L p \psi \, dz \tag{4.26}$$

are called, respectively, the generalised mass, stiffness and load. Equation 4.25 is an undamped SDOF equation of motion in terms of the generalised co-ordinate X, which can be solved by the methods presented in Chapter 2, with the full displacement solution then recovered from Equation 4.20. It must be remembered, of course, that this is an approximate method whose accuracy depends on the choice of the shape function.

We can now think about how to include damping. In principle, it would be possible to model the distributed damping within the beam and include it in the above analysis, but given the approach we have adopted to damping in previous chapters, it makes more sense to use a simplified global damping parameter such as the damping ratio ξ. Then, dividing through by m^*, we can write the damped equation of motion in the familiar form:

$$\ddot{X} + 2\xi \omega_n \dot{X} + \omega_n^2 X = \frac{p^*}{m^*} \tag{4.27}$$

Clearly, this process has much in common with the mode superposition procedure outlined in Section 4.2.3, the key differences being that only a single deformed shape is allowed, and that the shape function is a best guess rather than an analytical solution to a free vibration equation.

Example 4.4

Find the generalised mass and stiffness, and hence estimate the fundamental natural frequency, of a uniform beam of span L, mass per unit length m and flexural rigidity EI, based on the following shape functions (plotted in Figure 4.8):

$$\text{(a) } \psi = \sin \frac{\pi z}{L}, \quad \text{(b) } \psi = \frac{4z}{L}\left(1 - \frac{z}{L}\right), \quad \text{(c) } \psi = \sin \frac{2\pi z}{L}$$

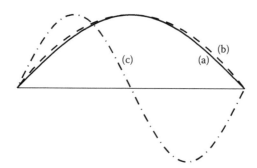

Figure 4.8 Shape functions for Example 4.4.

All that is required is the substitution of the shape functions into Equation 4.26 and some simple integration.

(a)

$$m^* = \int_0^L m \sin^2 \frac{\pi z}{L} \, dz = m \left[\frac{z}{2} - \frac{L}{4\pi} \sin \frac{2\pi z}{L} \right]_0^L = \frac{mL}{2}$$

$$k^* = \int_0^L EI \frac{\pi^4}{L^4} \sin^2 \frac{\pi z}{L} \, dz = EI \frac{\pi^4}{L^4} \left[\frac{z}{2} - \frac{L}{4\pi} \sin \frac{2\pi z}{L} \right]_0^L = \frac{\pi^4 EI}{2L^3} = \frac{48.7 EI}{L^3}$$

$$f_n = \frac{1}{2\pi} \sqrt{\frac{k^*}{m^*}} = \frac{\pi}{2} \sqrt{\frac{EI}{mL^4}}$$

In this case, the assumed shape function is identical to the fundamental mode shape of a simply supported beam, as calculated in Example 4.1. Unsurprisingly, the resulting natural frequency is exactly equal to that of the fundamental mode.

(b)

$$m^* = \int_0^L m \frac{4z}{L} \left(1 - \frac{z}{L} \right) dz = \frac{4m}{L} \left[\frac{z^2}{2} - \frac{z^3}{3L} \right]_0^L = \frac{2mL}{3}$$

$$k^* = \int_0^L EI \frac{64}{L^4} \, dz = EI \frac{64}{L^4} [z]_0^L = \frac{64EI}{L^3}$$

$$f_n = \frac{1}{2\pi} \sqrt{\frac{k^*}{m^*}} = \frac{\sqrt{24}}{\pi} \sqrt{\frac{EI}{mL^4}} = 1.559 \sqrt{\frac{EI}{mL^4}}$$

This time we have approximated the fundamental mode shape by a quadratic function, which is a little easier to manipulate algebraically, and to integrate. Although the two look broadly similar (Figure 4.8), the differences are sufficient to give values of modal mass and stiffness about 30% higher than those achieved using the exact mode shape. When calculating the natural frequency, these errors roughly cancel out – the frequency estimate in (b) is within 1% of the exact value. However, they may be more significant when calculating the response to an applied load.

(c)

$$m^* = \int_0^L m \sin^2 \frac{2\pi z}{L} \, dz = m \left[\frac{z}{2} - \frac{L}{8\pi} \sin \frac{4\pi z}{L} \right]_0^L = \frac{mL}{2}$$

$$k^* = \int_0^L EI \frac{16\pi^4}{L^4} \sin^2 \frac{2\pi z}{L} \, dz = EI \frac{16\pi^4}{L^4} \left[\frac{z}{2} - \frac{L}{8\pi} \sin \frac{4\pi z}{L} \right]_0^L = \frac{8\pi^4 EI}{L^3}$$

$$f_n = \frac{1}{2\pi}\sqrt{\frac{k^*}{m^*}} = 2\pi\sqrt{\frac{EI}{mL^4}}$$

The shape function used here is that of the second mode. This results in the same generalised mass as in (a) but in a much higher stiffness, and the natural frequency of the second mode is reproduced exactly. It is thus possible to derive an SDOF system representing any mode of vibration (not just the fundamental) by the use of an intelligently chosen shape function.

Lastly, note that the generalised displacement does not refer to the same point in each case: in (a) and (b) it is the midspan displacement (where the shape function is 1), whereas in (c) it is the displacement at the quarter-span point.

Example 4.5

A tapered chimney is modelled as a cantilever of length $L = 40$ m. The mass per unit height reduces linearly from $m_0 = 750$ kg/m at the base to 500 kg/m at the top, while the flexural rigidity reduces quadratically from $EI_0 = 2 \times 10^9$ Nm^2 at the base to 1×10^9 Nm^2 at the top. By choosing an appropriate shape function, estimate the vibration properties in the fundamental mode.

The mass and stiffness variations can be expressed algebraically as:

$$m(z) = m_0\left(1 - \frac{1}{3}\frac{z}{L}\right) \quad EI(z) = EI_0\left(1 - \frac{1}{2}\left(\frac{z}{L}\right)^2\right)$$

Although the chimney is not uniform, the fundamental mode shape would be expected to follow broadly the shape of the first cantilever mode shown in Figure 4.3. With the generalised co-ordinate defined as the cantilever tip displacement, the shape function must take a value of unity at $z = L$, and the boundary conditions can therefore be written as

$$\psi(0) = 0, \quad \psi'(0) = 0, \quad \psi(L) = 1, \quad \psi''(L) = 0$$

(The last boundary condition follows from the fact that the tip moment must be zero.) Given that both $m(z)$ and $EI(z)$ are polynomial, the mathematics will be greatly simplified if we also use a polynomial shape function, as opposed to sinusoidal, for instance. With four boundary conditions, it is possible to define a cubic, of the form:

$$\psi = Az^3 + Bz^2 + Cz + D$$

and substituting in the boundary conditions gives

$$\psi = \frac{3}{2}\frac{z^2}{L^2} - \frac{1}{2}\frac{z^3}{L^3}$$

We can now compute the generalised mass and stiffness as before. The integrals are a bit long-winded and only the results are shown here:

$$m^* = \int_0^L m\psi^2\,dz = \int_0^L m_0\left(1 - \frac{1}{3}\frac{z}{L}\right)\left(\frac{3}{2}\frac{z^2}{L^2} - \frac{1}{2}\frac{z^3}{L^3}\right)^2 dz = \frac{577}{3360}m_0L = 5152 \text{ kg}$$

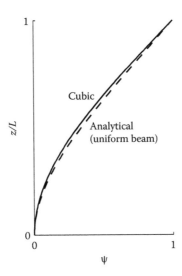

Figure 4.9 Shape functions for Example 4.5.

$$k^* = \int_0^L EI\psi''^2\,dz = \int_0^L EI_0\left(1-\frac{1}{2}\frac{z}{L}\right)\left(\frac{3}{L^2}\left(1-\frac{z}{L}\right)\right)^2 dz = \frac{147}{40}\frac{EI_0}{L^3} = 114.8\,\text{kN/m}$$

Hence, the natural frequency is

$$f_n = \frac{1}{2\pi}\sqrt{\frac{k^*}{m^*}} = 0.75\,\text{Hz}$$

Of course, performing an explicit free vibration analysis of this structure using the methods of Section 4.2 would be complex, due to the variable mass and stiffness. However, we can use the solution to the uniform cantilever case to do a quick reasonableness check on the above figures.

For the sake of simplicity, assume uniform values of m and EI obtained from the mid-height of the tapered structure, that is, $m = 625$ kg/m and $EI = 1.75 \times 10^9$ Nm². Using the appropriate formula in Table 4.1, this results in an estimate of f_n of 0.59 Hz. A comparison of the exact mode shape for a uniform beam and the cubic used for the tapered beam is shown in Figure 4.9. Clearly, there are some differences both in the mode shape and (rather more markedly) in the frequency estimate. However, both are in the same ballpark and it should be recalled that the analytical solution is itself an approximation.

4.3.2 Response to applied loads

Having found the generalised properties of a structure, calculating the dynamic response to an applied load then simply requires converting the load to its generalised form, performing an SDOF response analysis as in Chapter 2, and transforming the results back to the physical co-ordinate system.

Example 4.6

A uniform, simply supported beam has a span of $L = 10$ m, a mass per unit length (including supported elements that vibrate with the floor) of $m = 500$ kg/m and flexural rigidity $EI = 2 \times 10^8$ Nm². It supports one edge of a floor in such a way that forces from the floor are transmitted to the beam as a triangular distribution, as shown in Figure 4.10. Assuming 5% damping, calculate the response if a load is suddenly applied to the floor such that value of the peak load w on the beam increases sharply from 8 kN/m to 20 kN/m, as shown in Figure 4.10.

A simple bending analysis gives the static deflected shape due to a triangular load. Due to the nature of the load distribution, the mathematical function describing the shape has a discontinuity at midspan. For the left-hand half, it can be written as

$$x = \frac{wL^4}{960EI}(16y^5 - 40y^3 + 25y), \quad y = \frac{z}{L}, \quad 0 \le y \le 0.5$$

and of course the right-hand half will be the mirror image of this. We would expect the beam to adopt this shape under the initial load, and after the free vibrations caused by the sudden load change have been damped away. It therefore seems reasonable to adopt a shape function of a similar form. Normalising to a peak value of 1 at $z = L/2$:

$$\psi(z) = \frac{1}{8}(16y^5 - 40y^3 + 25y), \quad y = \frac{z}{L}, \quad 0 \le y \le 0.5$$

Using this shape function, the generalised mass and stiffness can be found in the usual way. We get

$$m^* = 2 \int_0^{L/2} m\psi^2 \, dz = 0.499mL = 2493 \text{ kg}$$

$$k^* = 2 \int_0^{L/2} EI\psi''^2 \, dz = 48.57 \frac{EI}{L^3} = 9.71 \text{ MN/m}$$

Comparing with Example 4.4(a), the generalised mass and stiffness are very similar to those obtained using a sinusoidal mode shape, suggesting that the fifth order polynomial is a close approximation to a sinusoid. The natural frequency is

$$\omega_n = \sqrt{\frac{k^*}{m^*}} = 62.4 \text{ rad/s}, \quad f_n = \frac{\omega_n}{2\pi} = 9.9 \text{ Hz}$$

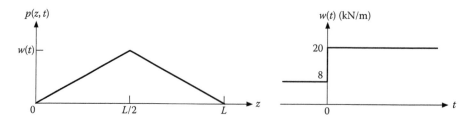

Figure 4.10 Dynamic load on a beam: triangular spatial distribution and step change in magnitude.

Now consider the loading. Over one half of the beam of the triangular load distribution can be represented by $p(z) = 2wy$. Initially there is a static load of $w = 8$ kN/m. At time $t = 0$ a step load of $w = 12$ kN/m is applied in addition; this part produces the dynamic response. The generalised load corresponding to $w = 12$ kN/m is

$$p^* = 2 \int_0^{L/2} p\psi\,dz = 2 \int_0^{1/2} 2wy.\frac{1}{8}(16y^5 - 40y^3 + 25y^2).L\,dy = 0.405wL = 48.6\text{ kN}$$

The response of an SDOF system to a step load is given by Equations 2.25 and 2.26. For our case, with $\xi = 0.05$, the solution can be written as

$$X(t) = \frac{p^*}{k^*}\left[1 - \frac{e^{-\xi\omega_n t}}{\sqrt{1-\xi^2}}\sin(\omega_d t + \phi)\right], \quad \phi = \tan^{-1}\frac{\sqrt{1-\xi^2}}{\xi} = 1.521\text{ rads}$$

and the full displacement solution is

$$x(z,\,t) = X(t).\psi(z)$$

Finally, the bending moment distribution is given by

$$M(z,t) = EIx''(z,t) = EIX(t)\psi''(z) = \frac{EIX(t)}{L^2}(40y^3 - 30y)$$

A selection of results is plotted in Figure 4.11. The top plot shows the midspan displacement (i.e. the generalised co-ordinate) as a function of time. The beam starts with a static displacement due to the permanent load of 8 kN/m, undergoes a transient vibration due to the step load and finally settles to a larger displacement under the total load of 20 kN/m. The largest displacement occurs after one half-cycle of vibration, at $t = 0.05$ s.

The second graph shows the displaced shapes of the beam prior to the step load, at the instant of peak displacement and at the end of the analysis, when the vibrations have all but ceased. Because of the choice of shape function for the dynamic analysis, all three curves have the same form and only the amplitude varies. We know from the static analysis that the initial and final displaced shapes will take this form, but it is possible that a slightly different shape will be achieved during the vibrations if a full dynamic analysis were undertaken.

Finally, the bending moments are plotted at the same three instants. The moments are proportional to the curvature of the shape function and so, again, it is possible that during the vibrations the exact distribution will be slightly different from that shown.

4.3.3 Response to a base motion

If a beam is subjected to a base motion x_g rather than an applied load, then, as with lumped-parameter systems, its total movement x can be expressed as the sum of the ground motion and a deflection y relative to the ground:

$$x(z,\,t) = x_g(t) + y(z,\,t) \tag{4.28}$$

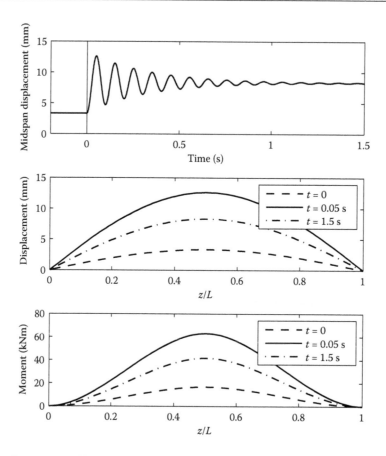

Figure 4.11 Displacements and bending moments for beam subjected to a step load.

Similarly to Equation 4.20, y can be written as the product of a shape function ψ and a generalised co-ordinate Y:

$$y(z, t) = Y(t).\psi(z) \tag{4.29}$$

Following a similar energy analysis to that of Section 4.3.1 then leads to

$$m^* \ddot{Y} + k^* Y = p^*_{\text{eff}} \tag{4.30}$$

where m^* and k^* are as defined in Equation 4.26 and the generalised effective force is

$$p^*_{\text{eff}} = -\ddot{x}_g \int_0^L m\psi \, dz \tag{4.31}$$

Note that Equation 4.30 is *not* exactly equivalent to the equation of motion for an SDOF system with a base motion (Equation 2.3, Chapter 2); if it were, we would expect the generalised effective force to be equal to the generalised mass times the ground acceleration. Instead, the transformation from distributed to equivalent SDOF system properties operates differently for the two parameters.

Example 4.7

Calculate the response of the chimney in Example 4.5 to the El Centro earthquake record (plotted in Figure 1.20, Chapter 1). Take the damping as 2% of critical.

In Example 4.5, using a cubic shape function, the generalised mass and stiffness were calculated as $m^* = 5152$ kg and $k^* = 114.8$ kN/m, giving a natural frequency of 0.75 Hz (or a period of 1.33 s). The effective force due to the earthquake is given by Equation 4.31:

$$p_{eff}^* = -\ddot{x}_g \int_0^L m\psi \, dz = -\ddot{x}_g \int_0^L m_0 \left(1 - \frac{1}{3}\frac{z}{L}\right)\left(\frac{3}{2}\frac{z^2}{L^2} - \frac{1}{2}\frac{z^3}{L^3}\right) dz$$

$$= -\frac{17}{60} m_0 L \ddot{x}_g = -8500 \ddot{x}_g$$

The response of the SDOF system to an irregular, time-varying load can be found using Duhamel's integral, as described in Section 2.4.6, Chapter 2. Applying this to our system gives the solution for the generalised co-ordinate Y, which is the displacement of the tip of the cantilever.

Figure 4.12 shows time histories of the effective force, and the resulting relative displacement and absolute acceleration (i.e. including the ground acceleration) at the tip

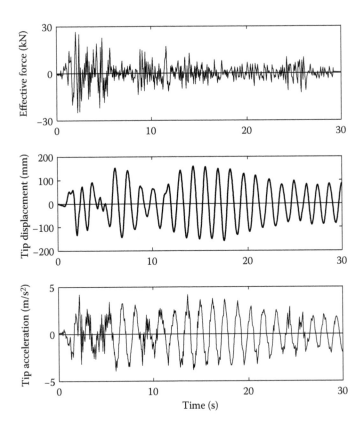

Figure 4.12 Tapered chimney under earthquake load: time histories of effective force, tip displacement (*relative* to ground) and tip acceleration (*absolute* – including ground motion).

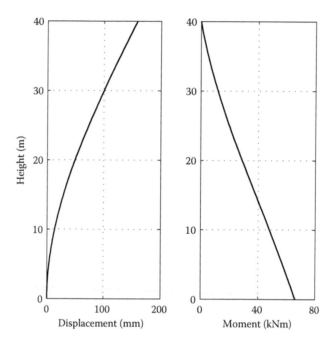

Figure 4.13 Tapered chimney under earthquake load: displaced shape and bending moment diagram at instant of maximum displacement ($t = 14.32$ s).

of the cantilever. The natural period of the structure is a little beyond the peak of the response spectrum for this earthquake (plotted in Figure 2.21) and the dynamic amplifications are not large. Interestingly, the largest motions are experienced in the region between 12 and 20 s, some time after the strongest part of the ground motion. Note also that, with relatively low damping, oscillations of the structure continue for some time, and are still quite strong at the end of the earthquake record ($t = 30$ s).

Having solved for the generalised co-ordinate, the full displacement can be recovered from Equation 4.29 and the bending moments can be found from:

$$M(z,t) = EI(z).y''(z,t) = EI(z)Y(t)\psi''(z) = Y(t).EI_0\left(1 - \frac{1}{2}\frac{z}{L}\right).\frac{3}{L^2}\left(1 - \frac{z}{L}\right)$$

In Figure 4.13, the deflected shape and bending moment diagram are plotted at the instant $t = 14.32$ s, when the tip deflection takes its maximum value.

KEY POINTS

- Dynamics of continuum elements such as beams and plates are governed by an equation of motion, which takes the form of a partial differential equation (PDE). This can be derived by relating the stiffness forces on an elemental mass to the inertia force as the mass accelerates.
- Solution of the PDE and imposition of appropriate boundary conditions yields natural frequencies and mode shapes. A continuous structure has an infinite number of DOFs and so also an infinite number of modes, though usually only the lowest ones are of interest.

- Response to dynamic loads can be achieved using a mode superposition approach similar to that used for lumped MDOF systems, but this can quickly become cumbersome for all but simple cases.
- The dynamics of continuous structures in one (usually the fundamental) mode of vibration can be conveniently reduced to an equivalent SDOF representation, by making a sensible estimate of the mode shape. This then allows the simpler SDOF solution methods of Chapter 2 to be applied. The accuracy with which the SDOF system replicates the true dynamic behaviour depends on how closely the estimated mode shape matches the true one. If the estimated shape is poorly chosen, very inaccurate results may be achieved.

TUTORIAL PROBLEMS

1. A uniform cantilever has length L, mass per unit length m, and flexural rigidity EI. It is rigidly fixed at $z = 0$ and free at $z = L$. Starting from the general solution in Equation 4.8, find expressions for the mode shapes in terms of the frequency parameter α. (The answers you are aiming for can be found in Table 4.1.)

2. The cantilever in Problem 1 is now propped at $z = L$, that is, it can rotate but not deflect at that point. Find new expressions for the mode shapes in terms of α and show that α, and hence the modal frequencies, are given by the roots of the equation:

$$\tan \alpha L - \tanh \alpha L = 0$$

(Again, the required solutions are in Table 4.1.)

3. The simply supported beam in Example 4.3 has properties: $L = 7.5$ m, $EI = 20$ MNm², $m = 1000$ kg/m, $\xi = 0.02$. It is now loaded at its centre by a concentrated, harmonically varying force of amplitude 2500 N and frequency 4 Hz. Find expressions for its displacement and bending moment distribution, as functions of both position and time, assuming the response is dominated by the fundamental mode. Explain why this is a reasonable assumption in this case.

4. The simply supported beam in Example 4.3/Problem 3 is now subjected to a distributed dynamic load that varies harmonically with time but, at any instant, is uniform at all points in the span. If the amplitude is 1000 N/m and the frequency is 3 Hz, find the displacement response of the beam in its first and third modes of vibration.

5. A uniform cantilever has mass per unit length m, flexural rigidity EI and length L. Ignoring damping, create an equivalent SDOF model based on the assumed shape function

$$\psi(z) = 1 - \cos\left(\frac{\pi z}{2L}\right)$$

and hence derive an approximate formula for the fundamental natural frequency.

6. A uniform simply supported beam has mass per unit length m, flexural rigidity EI and length L. A rigid, lumped mass $M = \alpha m L$ is attached to the beam at midspan and vibrates with it. Find the generalised mass and stiffness of the equivalent SDOF system

and hence estimate the fundamental natural frequency, assuming the following mode shapes:

(a) $\psi(z) = \sin\dfrac{\pi z}{L}$ $0 \le z \le L$

(b) $\psi(z) = \dfrac{3z}{L} - \dfrac{4z^3}{L^3}$ $0 \le z \le L/2$, symmetrical about $z = L/2$

Compare the natural frequency estimates for values of α in the range 0 to 100 and comment on the results.

7. We wish to create an equivalent SDOF model of a uniform propped cantilever of length L, fixed at $z = 0$ and pinned at $z = L$. Show that a polynomial shape function of the form

$\psi(z) = Az^4 + Bz^3 + Cz^2$

satisfies the relevant boundary conditions and find the constants A, B and C.

8. A uniform propped cantilever has span $L = 8.0$ m, mass per unit length $m = 250$ kg/m, flexural rigidity $EI = 5.0$ MNm^2, and is subjected to a uniformly distributed load of amplitude 2 kN/m, which oscillates harmonically at a frequency of 10 Hz. Using an equivalent SDOF model based on the shape function in Problem 7 and neglecting damping, estimate the natural frequency of the beam and the peak dynamic displacement. Comment on the likely accuracy of this solution.

9. A uniform, simply supported beam has mass per unit length m and span L. Calculate the generalised load on the equivalent SDOF system for the four load cases shown in Figure P4.9, using a half sine-wave shape function.

10. Evaluate the peak midspan displacement of the simply supported beam in Problem 9 to each of the four loadcases in Figure P4.9, using the following numerical values. Length $L = 10$ m, mass per unit length $m = 200$ kg/m, flexural rigidity $EI = 10^7$ Nm^2, 5% damping. Take the loading frequency to be 3 Hz in all cases. In (a) $w = 2$ kN/m, in (b) and (c) $F = 10$ kN, in (d) $X_g = 0.05$ m.

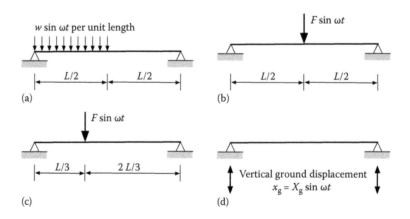

Figure P4.9

Chapter 5

Non-linear dynamics

5.1 INTRODUCTION

Chapters 1 to 4 have introduced the dynamics of *linear* structural systems. By this, we mean ones whose physical properties and overall geometry do not change significantly throughout the dynamic loading event. Linear systems can be described by relatively simple equations with constant coefficients. They obey the principle of superposition, allowing the response to be split into components that can then be combined by simple addition, and they scale straightforwardly, so that doubling the amplitude of the loading will double the response.

The behaviour of many structures under many loads is, or can reasonably be approximated as, linear. In structural dynamics, however, we are often designing against extreme events such as earthquakes or explosions. In these instances, it is usually impractical to design structures to respond in their linear elastic range. Instead, it is normal to accept some degree of damage short of collapse, and therefore some loss of linearity. We may then want to explore the non-linear behaviour of the damaged structure to ensure it retains overall stability and does not collapse. Dynamics engineers therefore need some understanding of non-linear behaviour, and of the very different analytical approaches that must be used to model it.

In this chapter, we will first consider in a little more depth what we mean by non-linearity, what forms it can take, and why conventional dynamic analysis techniques are no longer appropriate. We will then look at a couple of different time-stepping approaches to the solution of the non-linear equations of motion. In most of the chapter, we will concentrate on systems where the non-linearity is in the stiffness of the system, though we will also briefly consider other non-linearities.

This chapter should enable the reader to

- Understand the principal sources of non-linearity in structures,
- Create simplified models of non-linear stiffness elements and appreciate the level of approximation they involve,
- Recognise the key characteristics of non-linear dynamic behaviour,
- Understand the principles of different direct integration approaches to the calculation of non-linear dynamic response,
- Use direct integration methods within a finite element computer program with a clear understanding of their strengths and limitations,
- Make reasonable estimates of the stability and accuracy of a direct integration method,
- Analyse seismic structural response using ductility-modified response spectra.

5.2 NON-LINEARITY IN STRUCTURAL PROPERTIES

Non-linearity in structures can arise in two broad categories: through variation in the structural properties of mass, damping or stiffness, for instance through material yielding; or through large deformations of the structure, so that the normal small-displacement assumptions no longer apply. In this section, we will concentrate on the former case.

Taking the example of a single-degree-of-freedom (SDOF) system, we are now familiar with the idea that its dynamic behaviour is governed by an equation of the form

$$m\ddot{x} + c\dot{x} + kx = F(t) \tag{5.1}$$

In a linear system, the coefficients m, c and k are all constant, and Equation 5.1 is a linear, second-order differential equation, which has an explicit solution if $F(t)$ is a simple mathematical function, or otherwise can be solved by treating $F(t)$ as a series of impulses and summing the impulse responses. In a non-linear system, in principle, any of m, c or k could change during a dynamic loading event. Each possibility is considered below.

5.2.1 Variable mass

There are many dynamic systems where variations in mass are significant, such as rockets accelerating by expelling propellant. However, in vibrating structures the mass is unlikely to vary unless objects or parts of the structure fall off. The latter is an event that would generally be taken to constitute a catastrophic failure of the structure, such that it is unlikely one would want to continue the analysis beyond that point. For our purposes, therefore, m can normally be regarded as constant.

5.2.2 Non-proportional damping

Of more significance is the possibility of a non-linear damping term. Equation 5.1 is based on an idealisation that damping is provided by a linear viscous dashpot, which imposes a retarding force proportional to the velocity across it. Damping that deviates from this ideal is known as *non-proportional*.

As was discussed in Section 1.4.2 (Chapter 1), damping is in reality likely to come from a variety of sources, and at least some of it will probably be frictional. Friction is a highly non-linear phenomenon. A frictional contact between two surfaces requires a threshold force to be overcome before any relative motion occurs. Once this is reached, slip occurs at a constant force, regardless of the relative velocity across the contact.

Figure 5.1 shows the simplest mathematical model of a frictional contact. A slider is held against a rough surface with a normal force N, and a force F is applied along the surface. Friction generates a resisting force R, which cannot exceed μN, where μ is the coefficient of friction between the slider and surface. If F is increased from zero, initially R increases to match it and the object remains stationary. If F increases beyond μN, then R can no longer increase with it, equilibrium is lost and the object accelerates.

$$F \leq \mu N: \quad \ddot{x} = \frac{F - R}{m} = 0$$

$$F > \mu N: \quad \ddot{x} = \frac{F - \mu N}{m} \tag{5.2}$$

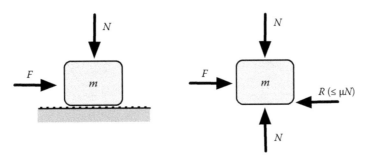

Figure 5.1 A block sliding on a frictional surface: applied forces (left) and free body diagram (right).

Once the mass is moving, different conditions pertain. Reducing F below μN will not immediately lead to a reduction in R or a regaining of static equilibrium. For this to happen, sliding must cease. So R remains at the value μN, causing the object to decelerate until its velocity relative to the surface is zero, at which point sliding ceases and will not recommence until F once again exceeds μN.

Figure 5.2 illustrates this process. In this case, the applied force is a sinusoid whose amplitude is steadily increased over time. At first, the force is insufficient to cause any movement, with the friction force exactly tracking the applied load. Then there are alternating cycles of slip in each direction, separated by periods of no slip. During slip, the friction force becomes constant and the sliding velocity of the block is non-zero. When slip ceases, the block comes to rest and the friction force returns to tracking the applied load. If the force amplitude were to become very large, slip would then occur repeatedly in opposite directions, with no rest period in between.

How would frictional behaviour affect structural vibrations? If the damper in our standard SDOF system were replaced by a frictional slider (Figure 5.3), then motion of the mass could only occur if the limiting friction were exceeded, so that the slider provided a force of constant magnitude in a direction opposing the velocity, for as long as the motion continued. The governing equation becomes:

$$m\ddot{x} + \mu N \,\mathrm{sgn}(\dot{x}) + kx = F(t) \tag{5.3}$$

where sgn() is the sign function (= +1 if the argument is positive, = −1 if the argument is negative). Even for free vibrations ($F(t) = 0$), solving the equation of motion now becomes difficult because we no longer have an ordinary differential equation with constant coefficients. However, we can assess the rate of decay of the vibrations by considering changes in energy over a cycle.

Figure 5.4 shows an arbitrarily drawn decaying vibration. At each of the points A, B and C the velocity, and therefore the kinetic energy, is zero. From A to B the velocity is negative and therefore the friction force is $+\mu N$; from B to C the velocity is positive and the friction force is $-\mu N$. Because of the changes in sign, we need to consider a half-cycle at a time. Between A and B the reduction in strain energy in the spring is caused by the work done against friction in the slider:

$$\frac{1}{2}k\left(x_A^2 - x_B^2\right) = \mu N(x_A - x_B) \quad \rightarrow \quad x_A + x_B = \frac{2\mu N}{k}$$

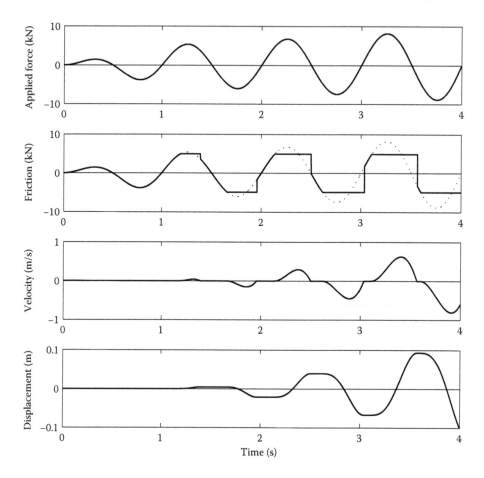

Figure 5.2 Response of a sliding block to a dynamic force. The applied force (top) is a ramped sinusoid. The response is split into regions where the frictional force opposing the motion balances the applied force and the block velocity is zero; and slip regions, where the block moves under the action of a constant friction force.

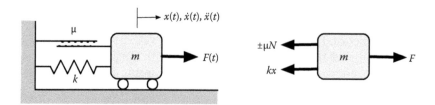

Figure 5.3 Single-degree-of-freedom system with frictional damping.

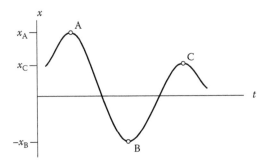

Figure 5.4 Amplitude variation over half-cycles of a decaying vibration.

Similarly between B and C:

$$\frac{1}{2}k\left(x_B^2 - x_C^2\right) = -\mu N(x_B - x_C) \;\rightarrow\; x_B + x_C = -\frac{2\mu N}{k}$$

Eliminating x_B from these equations gives the reduction in displacement amplitude over one period:

$$x_A - x_C = \frac{4\mu N}{k} \tag{5.4}$$

Since the right-hand side of Equation 5.4 is a constant, this implies a constant rate of reduction of displacement amplitude with time, so that a friction-damped free vibration will decay linearly, as opposed to the exponential decay we are used to seeing in a viscously damped system.

This behaviour can be seen in Figure 5.5, in which a friction-damped SDOF system released from an initial displacement experiences a linear decay in its free vibrations, until the magnitude of the forces falls below the slider's slip limit and the mass comes to rest. The end condition is worth noting: at each peak displacement, the mass comes to rest momentarily, before reversing the direction of its motion. On the final occasion, the spring force at this point is insufficient to generate motion in the slider, so the system stays put at a non-zero displacement, with the resulting spring force now in static equilibrium with an opposing damper force smaller than μN. The behaviour of the friction-damped system is markedly different from that of a viscous damper, which results in an exponential decay of free vibrations, in which it is the *ratio* between successive peaks that is constant, rather than the difference between them.

This difference can also be seen if we plot damping force as a function of the displacement of the mass (Figure 5.6). Since the frictional damper gives a constant force amplitude (either positive or negative), the resulting hysteresis loop is rectangular, whereas we saw that it is elliptical for a viscous damper (Figure 1.14, Chapter 1). Of course, Figure 5.6 also offers a way of converting a non-linear damping effect into an equivalent viscous damping c_{eq}, since the area inside the force-displacement loop gives the energy dissipation per cycle. Equating

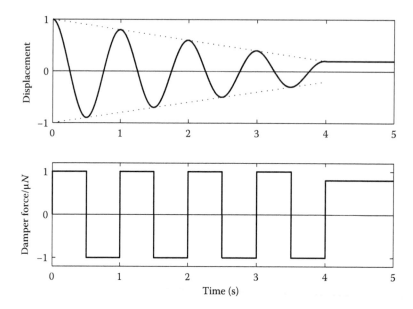

Figure 5.5 Free vibration of a frictionally damped SDOF system, showing (top) normalised displacement response and (bottom) damper force. Beyond 4 s there is static equilibrium between the spring and damper forces.

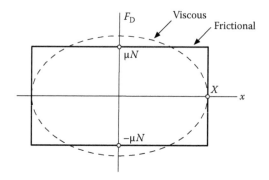

Figure 5.6 Force-displacement loop for a frictional damper undergoing a single vibration cycle.

the areas of the frictional and equivalent viscous loops (the latter was given in Equation 1.21, Chapter 1):

$$4\mu NX = \pi c_{eq}\omega X^2 \rightarrow c_{eq} = \frac{4\mu N}{\pi \omega X} \tag{5.5}$$

Clearly, this equivalence isn't straightforward to apply because the appropriate equivalent damping depends on both the amplitude and the frequency of the motion. This linearisation of the non-linear frictional damping is therefore likely always to be approximate.

Besides friction, other forms of damping non-linearity are possible. For example, in some applications purpose-built viscous dampers are added to structures to reduce their dynamic response. Such dampers are often not linear, with the retarding force instead being proportional to the velocity raised to a power:

$$F_D = \mathrm{sgn}(\dot{x}) c |\dot{x}|^{\alpha} \tag{5.6}$$

where the exponent α is normally less than 1. Note that in Equation 5.6 we take the modulus of the velocity before applying the exponent – this avoids any problems due to raising negative numbers to non-integer powers. We then need to multiply the result by the sign of the velocity to preserve the correct directionality of the damping force. Figure 5.7 shows how the shape of the damping force-displacement loop is affected by the value of α. When $\alpha = 1$, Equation 5.6 reduces to a linear viscous damper, giving an elliptical loop, as before. For smaller values of α, the loops become more square so that, for very small values of α, they become quite similar to the rectangular shape of the friction damper.

In multi-degree-of-freedom (MDOF) systems, non-proportional damping results in a coupling between modes of vibration, adding greatly to the complexity of structural response. However, in civil engineering applications it remains comparatively rare to engage in this level of complexity when certainty over the nature of the damping is difficult to achieve. We will therefore not pursue the possibility of non-linear damping in the analysis methods presented later in this chapter, though in principle it could be included in the formulation in a similar way to that presented for dealing with non-linear stiffness.

5.2.3 Non-linear stiffness

By far the most common form of non-linearity in structures arises due to variations in stiffness, usually due to yielding under extreme loads. The exact nature of the non-linearity will vary with the material, the member geometry and the structural form.

In frame structures, where the principal mode of structural action is bending, non-linearity generally takes the form of flexural hinges forming at the points of maximum bending moment.

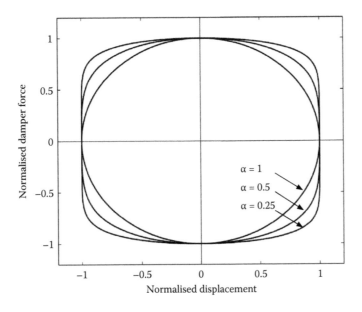

Figure 5.7 Force-displacement loops for non-linear viscous dampers.

As the moment increases, yielding commences at the extreme fibres of the section. Further increase in the moment causes yielding to spread inward, causing a gradual reduction in the rotational stiffness at the hinge location. Of course, yielding will also tend to spread lengthwise over a short span of beam around the point of maximum moment.

Some typical force-deflection characteristics of steel and concrete elements obtained in laboratory tests are shown in Figure 5.8. In each case, the components under test were subjected to cycles of alternating positive and negative load, with the amplitude increased steadily until failure. In Figure 5.8a the results of a test on a steel beam-column assemblage are plotted, while Figure 5.8b shows similar data for a test on a reinforced concrete bridge pier. In each case, the larger load-unload cycles cause repeated, alternating yielding in the positive and negative directions, so that the force-displacement curve follows a series of closed loops.

Behaviour of this sort is known as *hysteresis*, a word of Greek origin describing the phenomenon where a change in a physical property (in this case force or moment) leads or lags behind the property on which it depends (displacement or curvature). This results in the characteristic curves of Figure 5.8, in which the relation between the two properties comprises *hysteresis loops*, whose enclosed areas represent energy losses in the system.

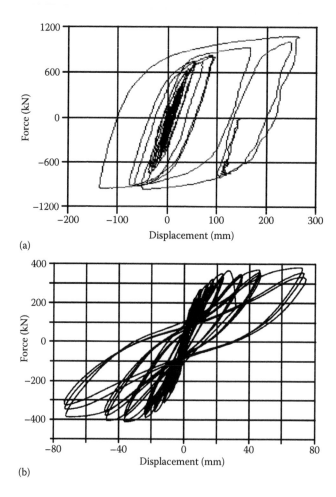

(a)

(b)

Figure 5.8 Typical force-displacement hysteresis loops from cyclic loading of: (a) a large steel beam-column joint and (b) a reinforced concrete bridge pier.

Some differences are evident between the steel and concrete elements. The steel curve in Figure 5.8a has relatively large, stable, open loops. Rather than displaying a sharp yield point on each cycle, the curves have a very rounded form in the bottom right and top left quadrants, a phenomenon known as the *Bauschinger effect*. For reinforced concrete, the cyclic behaviour is more complex – repeated loading tends to cause a deterioration in stiffness and strength due to concrete cracking and crushing, and local bond failure between concrete and reinforcing bars. As a result, the larger amplitude cycles have a flatter gradient and show a steady reduction in the peak force achieved even when the displacement amplitude is held constant. These phenomena are difficult to model as they can vary significantly depending on the element design; for instance, reinforced concrete elements designed to modern seismic codes are likely to show much less stiffness and strength degradation than older ones.

Even without including this level of detail in a mathematical model, hysteretic behaviour is complex to analyse because there is not a unique relationship between force and displacement. A given displacement may correspond to many different values of force (and vice versa), depending on the prior loading and the amount of plastic deformation.

For analysis purposes, we need to idealise the true behaviour illustrated in Figure 5.8, to make it more mathematically tractable. Below, we consider a couple of simple hysteresis models, which do not exactly describe real physical systems, but capture some of the key aspects of behaviour in a way that is not too mathematically complex.

5.2.3.1 Elastic-perfectly plastic hysteresis

The simplest form of stiffness non-linearity is the elastic-perfectly plastic (EPP) spring. This has an initial elastic stiffness k, a sharp yield point, and zero post-yield stiffness. For the purposes of this discussion, we will assume its behaviour is the same in extension and in compression. If the displacement x across the spring is increased monotonically from zero to beyond the yield point, the resisting force $R(x)$ takes the form shown in Figure 5.9a. This is often called the *skeleton curve* of the non-linear response, and can be expressed mathematically as

$$-x_y \leq x \leq x_y: \quad R(x) = kx$$
$$x > x_y: \quad R(x) = R_y = kx_y \quad\quad (5.7)$$
$$x < -x_y: \quad R(x) = -R_y = -kx_y$$

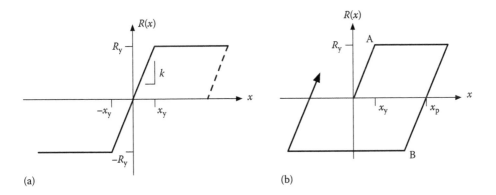

(a) (b)

Figure 5.9 Elastic-perfectly plastic stiffness behaviour: (a) skeleton curve, based on monotonic loading and (b) behaviour under cyclic load or deformation reversals.

If the spring is extended into the plastic range and then unloaded, the elastic part of its deformation will be recovered but the plastic displacement will remain, as shown by the dashed line in Figure 5.9a. The resulting end-point will then act as the origin for the forces and displacements in any subsequent loading.

In dynamics, repeated positive and negative cycles of deformation are common, as structures vibrate back and forth under oscillating loads. In this circumstance, the spring is likely to undergo successive plastic deformations in opposite directions (extension and compression) as shown in Figure 5.9b. Clearly this now becomes rather complex to analyse because the deformation at which yield occurs depends on the amount of plastic deformation up to that point. In the example shown, the first yield event (point A on the graph) occurs with the spring in tension at a positive displacement $x = x_y$. However, the second yield event (B), in compression, occurs when the spring has shortened by an amount $2x_y$ from its point of maximum extension, that is, at $x = x_p - x_y$. Therefore, to identify successive yield events in a complex loading cycle, we need to keep track of the accumulated plastic deformation x_p. Equation 5.7 can then be modified to take account of previous yield events, giving

$$
\begin{aligned}
-x_y \leq (x - x_p) \leq x_y\!: \quad & R(x) = k(x - x_p) \\
(x - x_p) > x_y\!: \quad & R(x) = R_y \\
(x - x_p) < -x_y\!: \quad & R(x) = -R_y
\end{aligned}
\tag{5.8}
$$

Figure 5.10 shows the output of a simple computer algorithm based on the above formulation, which calculates the force and plastic displacement of an EPP spring subjected to a prescribed set of displacements. The input displacement history and the resulting force history are shown, together with the force-displacement hysteresis behaviour. The prescribed displacement chosen is a sinusoid, whose amplitude is ramped up linearly from zero. Initially the behaviour is linear – the force response follows the imposed displacement – until yield occurs when the displacement first exceeds the yield threshold, and the force reaches a plateau. It is noticeable, however, that subsequent yield excursions do not simply occur when the yield displacement is exceeded because they are dependent on the prior history of plastic deformations as discussed above. In the time domain, therefore, the relationship between displacement and resisting force becomes increasingly less obvious with repeated yielding.

The hysteresis response shown in Figure 5.10b makes the behaviour clearer. Initially the force-displacement response simply cycles up and down an initial linear elastic curve passing through the origin. On the second cycle, a small amount of yielding occurs and the graph then comprises increasingly large, closed loops as the imposed displacement amplitude increases.

5.2.3.2 A bilinear stiffness model

A somewhat more realistic hysteresis model can be obtained by recognising that structural systems rarely have zero post-yield stiffness. Instead, strain hardening beyond the yield point results in a reduced but still positive stiffness; this latter stiffness may not be constant, but if we assume it is then we arrive at a mathematically manageable, bilinear behaviour.

Figure 5.11a shows the skeleton curve for this case, with initial stiffness k, a sharp yield point at a force R_y and post-yield stiffness k_2. If the spring is loaded beyond its yield point and then unloaded, it recovers elastically along a path parallel to the original gradient, and repeated cycling of the spring again results in closed hysteresis loops.

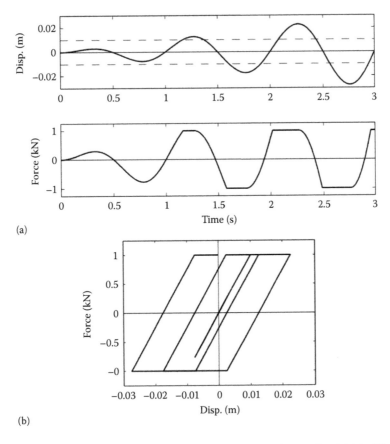

(a)

(b)

Figure 5.10 Elastic-perfectly plastic element under cyclic loading; the element has initial stiffness 100 kN/m and yield force 1 kN. (a) Imposed displacement (with yield displacement shown dashed) and resulting resistance force as functions of time and (b) force-displacement hysteresis behaviour.

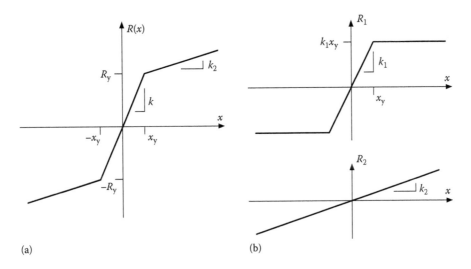

(a)

(b)

Figure 5.11 Bilinear stiffness model: (a) skeleton curve, (b) representation as the combination of an elastic-perfectly plastic component, R_1 and a linear component, R_2.

Hysteretic behaviour of this form can most easily be analysed by breaking it down into the sum of a linear component at the post-yield stiffness k_2, and an EPP component with initial stiffness k_1, such that $k_1 + k_2 = k$ (Figure 5.11b). R_2 is simply given by the linear relation

$$R_2 = k_2 x \qquad (5.9)$$

R_1 can be evaluated in a similar way to the EPP system described above. Comparing Figure 5.11b to Figure 5.9a, we see that the stiffness k must be replaced by k_1, and that the yield force of the EPP part is slightly lower than that of the full bilinear system, due to the subtraction of the R_2 component. Equation 5.8 can therefore be rewritten as

$$
\begin{aligned}
-x_y \le (x - x_p) \le x_y: &\quad R_1(x) = k_1(x - x_p) \\
(x - x_p) > x_y: &\quad R_1(x) = R_{y1} = k_1 x_y = R_y - k_2 x_y \\
(x - x_p) < -x_y: &\quad R_1(x) = -R_{y1}
\end{aligned}
\qquad (5.10)
$$

The total force corresponding to a displacement x is then given by

$$R = R_1 + R_2 \qquad (5.11)$$

Figure 5.12 shows the output from a bilinear stiffness computer model. The model parameters are the same as in the EPP model plotted in Figure 5.10, except that the post-yield stiffness k_2 is now taken as 10% of the initial elastic stiffness, instead of zero. The input displacement is also the same as in the earlier case. Clearly, the force generated in the spring element now continues to increase after yield, in proportion to the deformation. If, as is often the case, the post-yield deformations are large, then these force increases can be very significant. Another interesting point that can be observed from the hysteresis curve is that, after the initial yielding, subsequent yield events begin at a force other than the stated yield force – yield occurs at points on a line passing through the initial yield point and having a gradient k_2, and may occur at a larger or smaller force than the initial yield force, depending on the accumulated plastic strain at the time.

With the presence of a non-linear stiffness element, assuming the other components remain linear, our equation of motion now becomes

$$m\ddot{x} + c\dot{x} + R(x) = F(t) \qquad (5.12)$$

We shall explore ways of solving this equation in Section 5.3.

5.2.4 Geometric non-linearity

The overwhelming majority of structural analysis, both static and dynamic, relies on the assumption that deformations are small compared to the overall dimensions of the structure. The position and direction of applied forces therefore do not change significantly as the structure deforms. If this is not the case, the behaviour is said to be geometrically non-linear – that is, the non-linearity arises from changes in structural geometry rather than from the behaviour of the material, which may well still be within its linear elastic range. The basic principle can be illustrated using a simple pendulum.

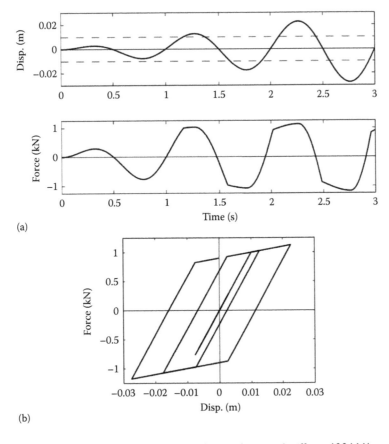

Figure 5.12 Bilinear element under cyclic loading; the element has initial stiffness 100 kN/m, post-yield stiffness 10 kN/m and yield force 1 kN. (a) Imposed displacement (with yield displacement shown dashed) and resulting resistance force as functions of time, (b) force-displacement hysteresis behaviour.

Example 5.1

Figure 5.13a shows a pendulum comprising a lumped mass m suspended from a light, rigid bar, swinging from a frictionless hinge. The pendulum is released from a large angle θ_0 and swings back and forth. At some instant, the bar makes an angle θ to the vertical.

For this system, it is more convenient to work in terms of angular motion than displacements. If the angular acceleration at any instant is $\ddot{\theta}$, then the tangential acceleration is $L\ddot{\theta}$. Resolving forces perpendicular to the bar:

$$mg\sin\theta = -mL\ddot{\theta}$$

which simplifies to

$$\ddot{\theta} + \frac{g}{L}\sin\theta = 0$$

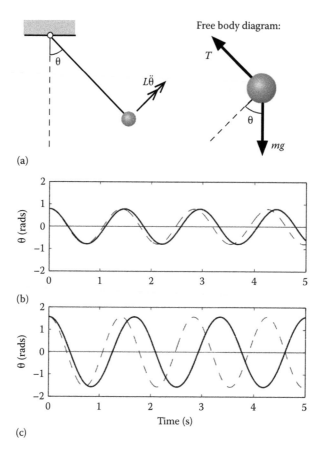

(a)

(b)

(c)

Figure 5.13 A pendulum undergoing large displacements: (a) shows the geometry definition and free body diagram. The pendulum angle is plotted as a function of time for cases with an initial angle of (b) π/4 rads and (c) π/2 rads. In each case, the solid line shows the geometrically non-linear solution and the dashed line the result obtained using small angle assumptions.

This is a non-linear differential equation, which cannot easily be solved analytically. In this example it is solved using the *central difference method*, a time-stepping method which will be set out later in Section 5.3.2.

Note that, if the pendulum motion were restricted to small displacements, we could use the small angle approximation $\sin \theta \approx \theta$, giving a linear equation of motion:

$$\ddot{\theta} + \frac{g}{L}\theta = 0$$

This is a standard ordinary differential equation of the sort we encountered in Chapter 2, and whose solution procedure is set out in Appendix B. The free oscillations computed from the non-linear equation are plotted in Figure 5.13b, and compared with the results of the linearised analysis (shown dashed). Results are shown for two different initial conditions: (b) $\theta_0 = \pi/4$, (c) $\theta_0 = \pi/2$. The larger the swing angle, the greater the deviation between the correct (non-linear) solution and the linearised one. In the linear analysis, the swing period is dependent only on the pendulum length and so is the same in both cases. Incorporating the non-linearity lengthens the swing period, the amount of lengthening increasing with the swing amplitude. It also causes the response to deviate somewhat from the perfect cosine shape of the linear solution, though this is quite hard to see in the plots shown.

In structures, geometric non-linearity may arise when the structure undergoes large deformations. Consider, for example, the cantilever beam of flexural rigidity EI subjected to a tip load P (Figure 5.14a). The curvature κ at any point is related to the bending moment M by

$$\kappa = \frac{M}{EI} \tag{5.13}$$

Using linear, small-deformation theory, κ is related to the deflection x at position z by

$$\kappa = \frac{d^2x}{dz^2} \tag{5.14}$$

Equations 5.13 and 5.14 are widely used for the calculation of beam deflections. Expressing the moment in terms of the applied load and integrating Equation 5.14 subject to appropriate boundary conditions, the static tip deflection $\delta = x(L)$ can be found as:

$$\delta = \frac{PL^3}{3EI} \tag{5.15}$$

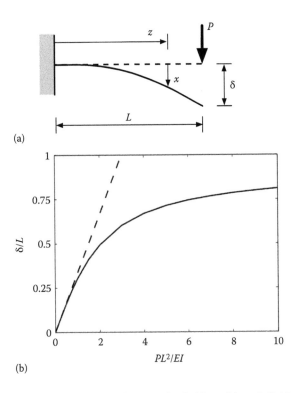

(a)

(b)

Figure 5.14 Static deflection of a cantilever under a tip load: (a) problem definition, (b) non-dimensional force-deflection plot showing geometrically non-linear analysis (solid line) compared to linear analysis based on small-deformation assumptions (dashed line).

However, Equation 5.14 is an approximation that is only accurate for small deformations. The mathematically exact expression for the curvature is

$$\kappa = \frac{\dfrac{d^2 x}{dz^2}}{\left[1 + (dx/dz)^2\right]^{3/2}} \quad (5.16)$$

The mathematics involved in solving this equation to find the deflection x are complex and not covered here. The resulting deflections are no longer linear functions of the load P, instead varying as shown in Figure 5.14b. It can be seen that the linear solution gives answers very close to the geometrically non-linear one up to deflections of about 20% of the beam length. Thereafter, the two solutions diverge, with the linear approximation significantly underestimating the stiffness of the beam at large deformations.

In Section 4.2.1, Chapter 4, the linear curvature definition, Equation 5.14, was used in the derivation of the governing equation for beam vibrations. If instead Equation 5.16 is used, the equation of motion for beam vibration becomes

$$\frac{\partial^2}{\partial z^2}\left(\frac{EI\dfrac{\partial^2 x}{\partial z^2}}{\left[1 + (\partial x/\partial z)^2\right]^{3/2}}\right) + m\frac{\partial^2 x}{\partial t^2} = p(z,t) \quad (5.17)$$

This is not amenable to direct solution and therefore we must resort to approximate numerical methods.

Problems involving geometric non-linearity are comparatively rare in structural engineering. In the remainder of this chapter, we will therefore focus on systems involving non-linear material behaviour.

5.2.5 Dynamic characteristics of non-linear systems

Before looking in detail at methods of solving non-linear equations of motion, we briefly introduce some of the fundamental principles of non-linear dynamic behaviour, by studying an undamped mass-spring system with bilinear stiffness, undergoing free vibrations at an initial amplitude sufficient to cause yielding.

Example 5.2

Consider a bilinear mass-spring system with the following properties: initial elastic stiffness 100 kN/m, yield force 1 kN, post-yield stiffness 40 kN/m, mass 2533 kg, zero viscous damping. Analyse the free vibration behaviour when the mass is given an initial velocity of 0.25 m/s.

Full analysis is not possible using the methods introduced thus far, but we can do some simple preliminary calculations. When vibrating within its elastic range, the system has a natural period of

$$T_n = 2\pi\sqrt{\frac{m}{k}} = 2\pi\sqrt{\frac{2533}{10^5}} = 1.0 \text{ s}$$

and its elastic displacement limit is

$$x_y = \frac{F_y}{k} = \frac{10^3}{10^5} = 0.01 \text{ m}$$

Plasticity will therefore occur if the displacement exceeds this value. In Figure 5.15, displacement and spring force time-histories have been calculated using the *central differ-ence method* (a non-linear time-stepping method that will be presented in Section 5.3.2). Clearly, the initial velocity is sufficient to cause a displacement of several times the yield value, so that the spring is extended well into its non-linear range. As a result, non-linear vibrations occur for much of the first 2 s of the analysis, followed by linear behaviour beyond 2 s, with the mass oscillating within its elastic range.

Evidence of the non-linearity is most obviously visible in the force plot, which shows several sharp kinks at points where the spring transitions between its two stiffness val-ues. Other characteristics of the non-linear behaviour are the facts that the force and displacement histories do not track each other proportionately during the early cycles of the response, and that the durations of each motion cycle are not equal – the first full cycle lasts approximately 1.4 s, compared to the elastic period of 1.0 s, which is achieved once the motion amplitude has reduced to the linear range.

Note also that the reduction in amplitude is itself an indicator of non-linearity. The sys-tem has no viscous damping and so the only cause of the reduction is the hysteretic energy absorption in the non-linear spring. The ratio between the second and first displacement peaks is 12.9/50.7 = 0.25 – to achieve this reduction by viscous damping would require a damping ratio of 21% of critical. Once the vibration returns to the linear regime, there is no damping and the amplitude of motion remains constant.

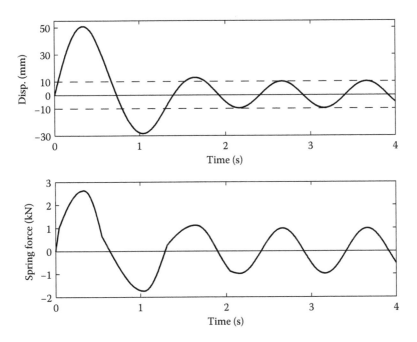

Figure 5.15 Free vibration of a bilinear SDOF system subject to an initial velocity: time histories of displace-ment of the mass (top – initial yield displacement shown dashed) and spring force (bottom).

To summarise, this example illustrates several key points about the dynamics of structures with stiffness non-linearities:

- Non-linearity breaks the simple, direct proportionality between spring force and extension.
- With a sharp yield point, sudden changes in gradient can occur in the force time-history.
- The system no longer has a fixed natural period, with the duration of vibration cycles varying due to the stiffness changes – indeed, the rate of motion may vary within a single cycle.
- As a general rule, a softening non-linearity such as occurs due to yielding will lengthen the vibration period.
- Spring hysteresis absorbs significant amounts of energy and so tends to damp the response.

5.3 TIME-STEPPING ALGORITHMS

5.3.1 The need for a different approach

Having described the characteristics of various non-linear structural systems, we will now focus on how to determine their dynamic response. This is going to require a fundamentally different method to linear systems, where we generally took a two-step approach:

1. Determine the intrinsic dynamic properties of the system – natural frequencies and (for MDOF systems) associated mode shapes.
2. Use these as the key building blocks in the determination of dynamic response to a given load – for instance, the level of dynamic amplification due to resonance may be deduced from the proximity of the structure's natural frequency to the loading frequency. MDOF systems may be analysed in terms of the response of each mode to the loading, with the total response then found by linear superposition.

Unfortunately, this strategy cannot be applied to non-linear systems. The natural frequencies and mode shapes are properties of a *linear* system, determined from the eigenvalues and eigenvectors of constant mass and stiffness matrices. If the stiffness varies with the amplitude of motion, then the stiffness matrix is no longer constant and its eigenvalues will continually change as the amplitude of the motion varies. Therefore, for a non-linear system, there is no single rate at which it prefers to oscillate and the concept of natural frequency is no longer meaningful. Of course, for a bilinear system such as those described in Section 5.2.3, it is possible to determine the natural frequencies and mode shapes of the initial linear system, before any yielding takes place. These can be useful to give a general idea of the likelihood of a severe resonant response, but if yielding is expected, then they cannot be used to determine the magnitude of that response. Even if one could calculate modal responses, they could not be simply combined, as the principle of linear superposition is not valid for non-linear systems.

Therefore, the normal approach to determining the response of a non-linear system is to solve the equation of motion at a series of closely spaced time steps. To reduce the number of unknowns to a manageable level, we make some assumption about the way the displacement varies with time. This allows us to express the velocities and accelerations in terms of the displacements at each time step, then solve for the displacements. Numerous different algorithms are available, based on different assumptions; we will look at only a couple of the simpler ones here, to illustrate the basic approach. These methods could also be applied to the analysis of linear systems, but in those cases, the methods presented in earlier chapters are generally preferred as they are simpler and more efficient.

The methods are presented here in terms of their application to SDOF systems but of course can be extended to the MDOF case straightforwardly; a few comments on this extension are offered in Section 5.3.5.

5.3.2 A simple algorithm: the central difference method

We will introduce time-stepping methods using a simple algorithm called the central difference method. This is not particularly widely used in practice but it is considered here as it gives us a relatively intuitive, easy-to-visualise way into the subject. Algorithms that are more sophisticated have many advantages but lack this clarity.

Suppose we knew the displacements of an SDOF system at a series of equally spaced times t_{i-1}, t_i, t_{i+1}, separated by a time step Δt (Figure 5.16). Velocity is rate of change of displacement, so we can write various approximations to the velocity at time t_i. For example, the forward difference approximation is t_{i-1}

$$\dot{x}_i = \frac{x_{i+1} - x_i}{\Delta t} \tag{5.18}$$

the backward difference approximation is

$$\dot{x}_i = \frac{x_i - x_{i-1}}{\Delta t} \tag{5.19}$$

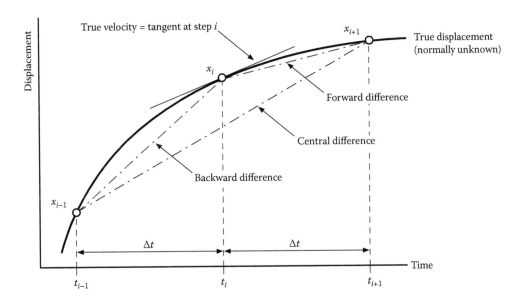

Figure 5.16 Finite difference approximations to a displacement-time curve.

and lastly the central difference approximation is

$$\dot{x}_i = \frac{x_{i+1} - x_{i-1}}{2\Delta t} \tag{5.20}$$

Obviously, Equation 5.18 is actually the average velocity looking forward over the next time step from t_i to t_{i+1}, Equation 5.19 is the average looking backward over the previous step and Equation 5.20 is the average velocity between t_{i-1} and t_{i+1}. The three estimates will tend to converge toward the same value as Δt is made smaller. For practical values of Δt, however, they will differ. Looking at Figure 5.16, the central difference formula appears to give the best estimate of the velocity at t_i, and indeed it can be shown mathematically to result in errors of a lower order; accuracy is discussed further in Section 5.3.3.

The approximate acceleration at time t_i can be found by taking the difference between the forward and backward velocity estimates in Equations 5.18 and 5.19:

$$\ddot{x}_i = \frac{1}{\Delta t}\left(\frac{x_{i+1} - x_i}{\Delta t} - \frac{x_i - x_{i-1}}{\Delta t}\right) = \frac{1}{\Delta t^2}(x_{i+1} - 2x_i + x_{i-1}) \tag{5.21}$$

Equations 5.20 and 5.21 are the central difference estimates of the velocity and acceleration. It can be shown quite easily that they are equivalent to assuming a quadratic variation in displacement (i.e. proportional to time squared) between times t_{i-1} and t_{i+1}, and therefore a linear variation in velocity and a constant acceleration over the same interval.

Now consider how these equations can be applied to the stepwise solution of a non-linear equation of motion. For simplicity, assume only the stiffness term is non-linear, as presented earlier in Equation 5.12:

$$m\ddot{x} + c\dot{x} + R(x) = F(t)$$

where $R(x)$ is the stiffness force, which may be a non-linear function of x. In the case of a linear system, the function $R(x)$ would simply equal kx where k is the stiffness of a linear spring. At time step i, this equation becomes

$$m\ddot{x}_i + c\dot{x}_i + R_i = F_i \tag{5.22}$$

If we substitute for $\dot{x}_i$ and $\ddot{x}_i$ from Equations 5.20 and 5.21, and rearrange, we get

$$\left[\frac{m}{\Delta t^2} + \frac{c}{2\Delta t}\right]x_{i+1} = F_i - R_i + \frac{2m}{\Delta t^2}x_i - \left[\frac{m}{\Delta t^2} - \frac{c}{2\Delta t}\right]x_{i-1} \tag{5.23}$$

This allows the displacement at step $i + 1$ to be determined based on the displacements at previous times.

Application of Equation 5.23 is straightforward, except at start-up, because the displacements at two previous times are needed; to determine x_1 would require knowledge not only of the initial displacement x_0 (which would normally be a known initial condition), but also of x_{-1}, which does not exist. We can get around this by applying Equations 5.20 and 5.21 at step zero:

$$\dot{x}_0 = \frac{x_1 - x_{-1}}{2\Delta t} \tag{5.24}$$

$$\ddot{x}_0 = \frac{1}{\Delta t^2}(x_1 - 2x_0 + x_{-1}) \tag{5.25}$$

then, eliminating x_{-1} from Equations 5.24 and 5.25 gives

$$x_1 = x_0 + \dot{x}_0 \Delta t + \ddot{x}_0 \frac{\Delta t^2}{2} \tag{5.26}$$

We can now formulate a step-by-step procedure to evaluate the dynamic response of the system. It is assumed that the forcing function F_i is known for all i, and that the stiffness force function R is fully defined, so that R_i can be determined directly from the current x_i and (in the case of a hysteretic model) its previous values.

1. Start from known initial conditions x_0, $\dot{x}_0$.
2. Find R_0 from the non-linear force-displacement relation.
3. Find $\ddot{x}_0$ by applying the equation of motion (Equation 5.21), at $i = 0$.
4. Find x_1 from the start-up equation (Equation 5.26).
5. For $i = 1, 2, 3...$
   a. Calculate R_i from non-linear force-displacement relation.
   b. Hence, find new displacement x_{i+1} from Equation 5.23.
   c. If desired, find velocity $\dot{x}_i$ from Equation 5.20.
   d. If desired, find acceleration $\ddot{x}_i$ from Equation 5.22. Note that the acceleration could also be found by back substituting into Equation 5.21, but using Equation 5.22 is likely to be more accurate because it ensures that the overall equation of motion is satisfied at each time step.

The central difference method (CDM) is an example of an *explicit* time-stepping method, meaning that the displacement at step $i + 1$ is determined on the basis of the state of the system at step i (Equation 5.23). Explicit schemes are generally valued for their ease of implementation, but they can require very short time steps to ensure stability, as discussed later.

Example 5.3

In this example, we calculate the response of a non-linear SDOF system to a dynamically varying applied force using the CDM. The system properties are

- Mass = 2000 kg, initial stiffness = 100 kN/m, hence initial period = 0.89 s,
- Viscous damping ratio (based on the initial elastic properties) = 0.05,
- Spring yield force = 1 kN, hence initial yield extension = 10 mm,
- Post-yield stiffness = 10 kN/m.

The system starts from rest and is loaded by the forcing function in Figure 5.17a, comprising four sinusoids of period 1 s and with increasing amplitudes of 0.25, 0.5, 1 and

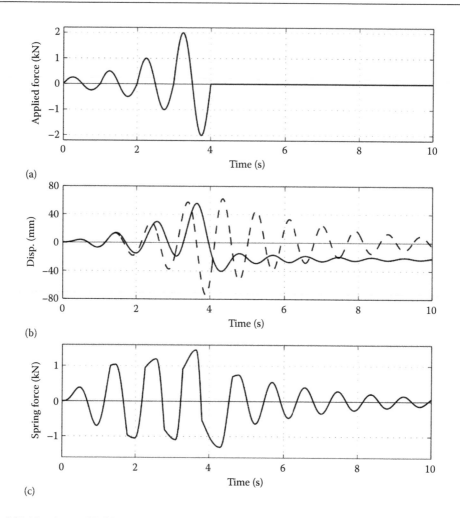

Figure 5.17 Non-linear SDOF system analysed using central difference method: (a) applied load history, (b) calculated displacement time history (with output from linear analysis shown dashed) and (c) time history of non-linear spring force.

2 kN. The response is calculated using a simple computer program implementing the central difference method as outlined above, with a time step Δt of 0.005 s.

Figure 5.17b and c show the resulting displacement of the system, and the resisting force in the spring, as functions of time. For comparison, Figure 5.17b also shows by a dashed line the output from a CDM analysis of a non-yielding structure, all other properties being the same.

At first, the forces are small enough that the structure responds linearly. There is near-resonance between the forcing and the elastic natural frequency of the structure, causing amplification of the response. Early in the second cycle of motion, the elastic limit is exceeded – this can be clearly identified by the sharp change in the gradient of the spring force time history and by the divergence of the displacement from that of the linear system. Thereafter, the system undergoes several cycles of non-linear hysteretic response. Compared to the linear system, the yielding structure oscillates at a longer period and displays a less symmetric response, with the negative peaks reduced in amplitude.

When the loading ceases at 4 s, the structure continues to oscillate with non-linear excursions for around a further second, until the hysteresis and the intrinsic viscous

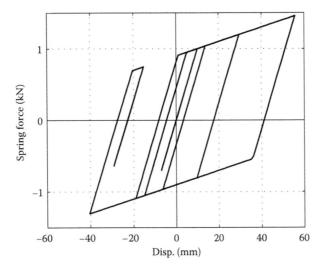

Figure 5.18 Non-linear SDOF system analysed using central difference method: spring hysteresis.

damping cause the amplitude to reduce to within the linear range. From around 5 s onward, the motion is viscously damped free vibration. However, the accumulated plastic deformation at the end of the last non-linear excursion is non-zero, with the result that these final oscillations are centred about a residual displacement of –22 mm.

The behaviour can also be understood from the hysteresis plot of Figure 5.18. Starting from the origin, the spring initially remains within the linear elastic regime, before following increasingly large hysteresis loops as the load amplitude increases. In the latter part of the response, it returns to cycles of linear elastic response (i.e. with no hysteresis loop), but this time centred on a point that crosses the zero force axis at –22 mm.

Example 5.4

We now evaluate the response of the same structural system as in Example 5.3, but this time to an earthquake input. We again use the El Centro earthquake ground motion, introduced in Figure 1.20 (Chapter 1). For comparison, we will also determine the response of a linear system with otherwise identical properties.

We first consider the linear system. This is solved using the CDM; because the earthquake record is digitised at steps of 0.02 s, this time step is used in the analysis. Figure 5.19 shows the displacement of the structure relative to the ground, and its absolute acceleration (i.e. the acceleration relative to the ground plus the ground acceleration). We also solved this problem in Example 2.11 in Chapter 2, using Duhamel's integral – as we would hope, we find that the same results are achieved by the CDM. As we are now used to seeing for elastic systems, there are signs of a resonant build-up of motion, and of a dominant natural period of vibration throughout the response. We also see that the relative displacement (which is proportional to the spring force) and absolute acceleration (proportional to the inertia force) have a similar form but opposite sign, so that the dynamic spring and inertia forces balance.

If we now repeat the analysis introducing the non-linearity, we see a very different form of response (Figure 5.20). The displacements are of a similar order of magnitude to the linear system, though slightly smaller – the maximum absolute value is 75 mm, compared to 110 mm for the linear system. However, there is no resonant build-up of response, and the rate of motion during the early, strongest part of the earthquake is

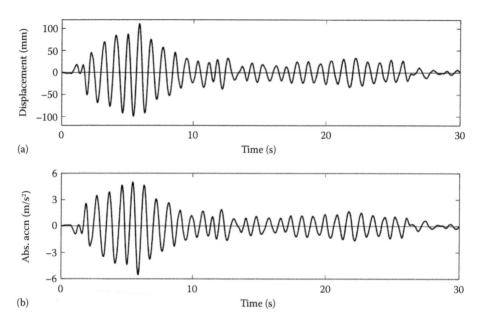

Figure 5.19 Response of linear SDOF system to El Centro earthquake, determined using the CDM: (a) relative displacement and (b) absolute acceleration.

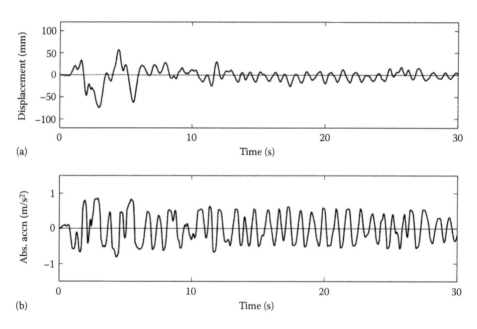

Figure 5.20 Response of a non-linear SDOF system to El Centro earthquake, determined using the CDM: (a) relative displacement and (b) absolute acceleration. (System is identical to that in Figure 5.19 except for the introduction of a yield point at a spring force of 1 kN, with post-yield stiffness equal to 10% of the elastic stiffness.)

much slower than in the linear case. Because of the yielding, the spring force corresponding to this displacement (not plotted) will no longer be directly proportional to the displacement.

The differences are even more obvious when we look at the accelerations. These are smaller than in the linear case by a factor of more than 5. Note that there is still a direct proportionality between the inertia force and the absolute acceleration, and that we would still expect the spring and inertia forces to balance. Since the size of the spring force is limited by yielding, the inertia force is similarly reduced, corresponding to lower accelerations.

5.3.3 Stability, accuracy and time step

One of the key decisions in implementing any direct integration algorithm is the choice of time step. For computational efficiency, we generally wish to avoid extremely short time steps. However, if we lengthen the time step excessively we will lose accuracy and may also run the risk of instability. It is worth defining and distinguishing between these two terms. By *accuracy* we mean deviation of the computed solution from the true response of the structure – this will generally be unknown, but we can nevertheless make an assessment by comparing algorithm performance for some reference cases where exact solutions are available, and by following some simple, common-sense rules. *Instability* occurs when errors accumulate or multiply from step to step, so that the computed solution grows without bound as the analysis proceeds.

5.3.3.1 Accuracy

First, consider accuracy. The acceptable level of accuracy of any numerical method is a matter of judgement by the user, and will need to take account of the sensitivity of the structure being designed and the level of uncertainty of other aspects of the analysis.

In practice, quantifying the accuracy of a particular numerical method is difficult. This is partly because the error arises from two sources – a *discretisation* error due to the inherent approximation in the time-stepping scheme, and a *round-off* error arising from the computer rounding of decimal quantities. When considering the accuracy of a time-stepping scheme, it is generally the discretisation error that is the main focus, and this is normally expressed in terms of its order of magnitude in relation to the time step.

For the example of the CDM, we can come up with an accuracy estimate by using Taylor's theorem. This states that if a function $x(t)$ is differentiable many times, then its value at a time increment Δt from a time t_0 can be expressed as an infinite series in terms of the values of the function and its derivatives at t_0:

$$x(t_0 + \Delta t) = x(t_0) + \Delta t \dot{x}(t_0) + \frac{\Delta t^2}{2!}\ddot{x}(t_0) + \frac{\Delta t^3}{3!}x^{(3)}(t_0) + \frac{\Delta t^4}{4!}x^{(4)}(t_0) + \ldots \quad (5.27)$$

Here, the superscripted numbers in parentheses denote the higher time derivatives, for example, $x^{(3)}(t_0)$ is the third derivative of x with respect to t, evaluated at time t_0. Similarly:

$$x(t_0 - \Delta t) = x(t_0) - \Delta t \dot{x}(t_0) + \frac{\Delta t^2}{2!}\ddot{x}(t_0) - \frac{\Delta t^3}{3!}x^{(3)}(t_0) + \frac{\Delta t^4}{4!}x^{(4)}(t_0) - \ldots \quad (5.28)$$

If we subtract Equation 5.28 from Equation 5.27 and rearrange, we get an expression for the velocity at time t_0:

$$\dot{x}(t_0) = \frac{x(t_0 + \Delta t) - x(t_0 - \Delta t)}{2\Delta t} - \frac{\Delta t^2}{3!} x^{(3)}(t_0) - \dots \tag{5.29}$$

Alternatively, adding and rearranging the same equations gives the following expression for the acceleration:

$$\ddot{x}(t_0) = \frac{x(t_0 + \Delta t) - 2x(t_0) + x(t_0 - \Delta t)}{\Delta t^2} - \frac{2\Delta t^2}{4!} x^{(4)}(t_0) - \dots \tag{5.30}$$

If we now compare these equations with the central difference equations derived earlier (Equations 5.20 and 5.21), we immediately see that the first terms on the right-hand side are the central difference approximations to the velocity and acceleration. Therefore:

$$\dot{x}(t_0) = \dot{x}_{CDM}(t_0) - \frac{t^2}{3!} x^{(3)}(t_0) - \dots = \dot{x}_{CDM}(t_0) + O(\Delta t^2) \tag{5.31}$$

$$\ddot{x}(t_0) = \ddot{x}_{CDM}(t_0) - \frac{2\Delta t^2}{4!} x^{(4)}(t_0) - \dots = \ddot{x}_{CDM}(t_0) + O(\Delta t^2) \tag{5.32}$$

in which $O(\Delta t^2)$ means terms of order Δt^2, that is, terms involving Δt raised to the power of 2 or higher.

This analysis provides an alternative derivation of the CDM to the graphical method used earlier and also demonstrates that the method has *second-order accuracy*. That is, as we reduce the time step, we expect the discretisation error to reduce proportionally to the square of the time step so that, for example, halving the step size would reduce the error by a factor of 4.

Of course, the actual magnitude of the error is also related to the derivatives of the function being modelled, so that, for a given time step, one might expect larger errors when the response of the system being analysed varies rapidly. For this reason, it is impossible to give absolute guidance on accuracy as a function of time step, independent of a consideration of the problem being modelled. When performing an analysis, it may be necessary to try several different step lengths, reducing until the resulting change in response becomes acceptably small. In doing this, it can be helpful to split the error into two components as illustrated in Figure 5.21, which shows in schematic form how the results of a time-stepping scheme can differ from an exact, analytical solution. The difference can most neatly be summarised in terms of a proportionate amplitude error e_{amp}:

$$e_{amp} = \left| \frac{X_{di} - X_{an}}{X_{an}} \right| \tag{5.33}$$

and a period error e_{per}:

$$e_{per} = \left| \frac{T_{di} - T_{an}}{T_{an}} \right| \tag{5.34}$$

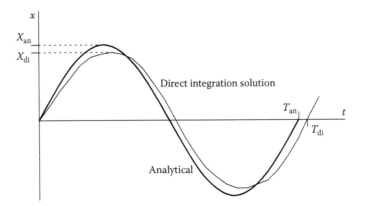

Figure 5.21 Schematic comparison between an exact, analytical solution and its estimation by a numerical direct integration scheme.

where the amplitude and period parameters are as defined in Figure 5.21. Note that this characterisation is far more meaningful than focusing on the error in signal magnitude at any instant. As can be seen in Figure 5.21, even a small period error can result in quite a large instantaneous difference in magnitude on the steep part of the response, but it would be misleading to interpret this as a large error because the two signals shown are in fact very similar to each other and have similar peak values.

One other simple consideration, regardless of the time-stepping method used, is how frequently a time-varying signal needs to be sampled in order to capture the peaks with reasonable accuracy. There is a high chance that a peak response will occur somewhere between two sampling points, leading to an underestimation of the true peak amplitude. As an illustration, Figure 5.22 shows a simple sine wave sampled at 20 and 40 points per cycle. Both map the curve quite well, but the higher sampling rate clearly gives a better fit around the peak. In the worst case, when the peak happens to lie midway between two sampling points, sampling at 20 points per cycle underestimates the true peak value by 1.23%, while for the 40 points per cycle case the error is only 0.31%. A good guide, therefore, is that the peaks in the signal can be accurately captured if the time step is no greater than one-fortieth of the shortest period of interest.

5.3.3.2 Stability

Stability is most readily assessed by recasting the time-stepping equations into a matrix recurrence relation. A rigorous treatment is difficult for non-linear systems, so we will

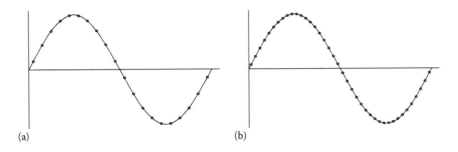

(a)   (b)

Figure 5.22 Discretisation of a sine wave by (a) 20- and (b) 40-equally-spaced points per cycle.

consider the example of the CDM applied to a linear system, that is, one where the spring resistance force $R = kx$. With this simplification, Equation 5.23 becomes

$$\left[\frac{m}{\Delta t^2} + \frac{c}{2\Delta t}\right]x_{i+1} = \left[\frac{2m}{\Delta t^2} - k\right]x_i - \left[\frac{m}{\Delta t^2} - \frac{c}{2\Delta t}\right]x_{i-1} + F_i$$

which can be rearranged to give

$$x_{i+1} = px_i - qx_{i-1} + rF_i \qquad (5.35)$$

where the coefficients p, q and r are functions of the system properties and the time step:

$$p = \frac{\left[\dfrac{2m}{\Delta t^2} - k\right]}{\left[\dfrac{m}{\Delta t^2} + \dfrac{c}{2\Delta t}\right]}, \; q = \frac{\left[\dfrac{m}{\Delta t^2} - \dfrac{c}{2\Delta t}\right]}{\left[\dfrac{m}{\Delta t^2} + \dfrac{c}{2\Delta t}\right]}, \; r = \frac{1}{\left[\dfrac{m}{\Delta t^2} + \dfrac{c}{2\Delta t}\right]} \qquad (5.36)$$

If we also introduce the obvious statement that $x_i = x_i$, we can combine this with Equation 5.35 to form the matrix equation:

$$\begin{bmatrix} x_{i+1} \\ x_i \end{bmatrix} = \begin{bmatrix} p & -q \\ 1 & 0 \end{bmatrix}\begin{bmatrix} x_i \\ x_{i-1} \end{bmatrix} + \begin{bmatrix} r \\ 0 \end{bmatrix}F_i \qquad (5.37)$$

which we can express more concisely as

$$\hat{\mathbf{x}}_{i+1} = \mathbf{A}\hat{\mathbf{x}}_i + \mathbf{b}F_i \qquad (5.38)$$

Suppose now that, at each time step, the approximation inherent in our algorithm generates an error term $\mathbf{e}_i$. In other words, the computed displacements at the end of the time step differ from the (usually unknown) exact solution $\mathbf{s}_i$:

$$\hat{\mathbf{x}}_i = \mathbf{s}_i + \mathbf{e}_i \qquad (5.39)$$

Substituting into Equation 5.38,

$$\hat{\mathbf{x}}_{i+1} = \mathbf{A}(\mathbf{s}_i + \mathbf{e}_i) + \mathbf{b}F_i \qquad (5.40)$$

If the time-stepping method were exact, then we would have simply

$$\mathbf{s}_{i+1} = \mathbf{A}\mathbf{s}_i + \mathbf{b}F_i \qquad (5.41)$$

Subtracting Equation 5.41 from Equation 5.40 gives the error in step $i + 1$ due to an error in step i:

$$\mathbf{e}_{i+1} = \mathbf{A}\mathbf{e}_i \qquad (5.42)$$

The stability of the system then depends on the properties of the amplification matrix **A**. It can be seen intuitively that, if the terms in **A** are small, an error introduced in a particular time step will be reduced in subsequent time steps and will quickly become insignificant. (Of course, this alone does not guarantee accuracy because new errors are likely to be introduced at each step, and these must be assessed and controlled as discussed previously.) On the other hand, if the terms in **A** are large, even very small errors in the response parameters will tend to be amplified with every time step, resulting in an unstable response. A more formal mathematical analysis reveals that stability can be guaranteed if the *spectral radius* of matrix **A** (i.e. the magnitude of its largest eigenvalue) is less than 1. Following this analysis through for the CDM results in a limit on the time step in terms of the natural period of the system:

$$\Delta t < 2\sqrt{\frac{m}{k}} \ \rightarrow \ \Delta t < \frac{T_n}{\pi} \tag{5.43}$$

For an SDOF system, this limit is unlikely to be problematic – we have already seen that we are likely to require a shorter time step than this in order to track a dynamic response with sufficient accuracy. However, the situation will become more complex when we look at analysis of MDOF systems in Section 5.3.5.

5.3.4 A more complex algorithm: the Newmark method

To illustrate the variety of time-stepping methods available, we will also consider a rather more complex one – the Newmark method. This method is based on the following equations describing how the velocity and displacement vary between time steps:

$$\dot{x}_{i+1} = \dot{x}_i + \Delta t\left[(1-\gamma)\ddot{x}_i + \gamma\ddot{x}_{i+1}\right] \tag{5.44}$$

$$x_{i+1} = x_i + \Delta t\,\dot{x}_i + \Delta t^2\left[(0.5-\beta)\ddot{x}_i + \beta\ddot{x}_{i+1}\right] \tag{5.45}$$

where β and γ are user-defined parameters between 0 and 1. In fact, Equations 5.44 and 5.45 define a family of methods, whose meaning and properties vary with these two parameters. For instance, if $\beta = 0.25$ and $\gamma = 0.5$, then the equations are those that would be obtained by assuming a constant acceleration over the time step, equal to the mean of the accelerations at steps i and $i + 1$. Alternatively, taking $\beta = 0.167$ and $\gamma = 0.5$ is equivalent to assuming the acceleration varies linearly over the time step.

Equations 5.44 and 5.45 can be rearranged to give the following expressions for the velocity and acceleration at step $i + 1$:

$$\dot{x}_{i+1} = \frac{\gamma}{\beta\Delta t}(x_{i+1}-x_i) + \left(1-\frac{\gamma}{\beta}\right)\dot{x}_i + \Delta t\left(1-\frac{\gamma}{2\beta}\right)\ddot{x}_i \tag{5.46}$$

$$\ddot{x}_{i+1} = \frac{1}{\beta\Delta t^2}(x_{i+1}-x_i) - \frac{1}{\beta\Delta t}\dot{x}_i + \left(1-\frac{1}{2\beta}\right)\ddot{x}_i \tag{5.47}$$

Now consider the equation of motion at time step $i + 1$:

$$m\ddot{x}_{i+1} + c\dot{x}_{i+1} + R_{i+1} = F_{i+1} \qquad (5.48)$$

Substituting for the velocity and acceleration from Equations 5.46 and 5.47 and rearranging gives

$$\left(\frac{m}{\beta \Delta t^2} + \frac{\gamma c}{\beta \Delta t}\right)x_{i+1} = F_{i+1} - R_{i+1} + \left(\frac{m}{\beta \Delta t^2} + \frac{\gamma c}{\beta \Delta t}\right)x_i + \left(\frac{m}{\beta \Delta t} + \frac{\gamma c}{\beta} - c\right)\dot{x}_i$$

$$+ \Delta t \left(\frac{m}{2\beta} - m + \frac{\gamma c}{2\beta} - c\right)\ddot{x}_i \qquad (5.49)$$

With known initial conditions, Equation 5.49 can, in principle, be used to calculate displacements at successive time steps. There is a problem, however; the stiffness force R_{i+1} is a function of x_{i+1}, which is unknown at the start of the time step. In the case of a linear elastic system of stiffness k, we know that $R_{i+1} = kx_{i+1}$. We can substitute this into Equation 5.49, take the term kx_{i+1} over to the left-hand side of the equation, and thus solve directly for x_{i+1}. However, for a non-linear system, the relation between R and x is more complex. It therefore becomes necessary to use an iterative approach, in which we come up with an initial estimate of x_{i+1}, calculate the corresponding R_{i+1} from the known, non-linear force-displacement relation, and then use this estimate of R_{i+1} in Equation 5.49 to get an improved estimate of x_{i+1}. We can continue iterating in this way until successive estimates of x_{i+1} do not change significantly.

Unlike the CDM, the Newmark approach requires no special start-up procedure. Using the superscript k to define the iteration number within each time step, a suitable solution approach would be:

1. Start from known initial conditions x_0, $\dot{x}_0$.
2. Find R_0 from non-linear force-displacement relation.
3. Find $\ddot{x}_0$ by applying the equation of motion (Equation 5.48), at $i = 0$.
4. For $i = 0, 1, 2, 3...$
   a. Let $R_{i+1}^0 = R_i$.
   b. Hence, find x_{i+1}^0 from Equation 5.49.
   c. For $k = 1, 2, 3...$
      i. Estimate R_{i+1}^k from force-displacement relation, using x_{i+1}^{k-1}.
      ii. Hence, find x_{i+1}^k from Equation 5.49.
      iii. Repeat until successive estimates of x_{i+1} (and R_{i+1}) cease to change significantly.
   d. If desired, velocities and accelerations at each step can be found by back substituting into Equations 5.46 and 5.47.

The Newmark method is an example of an *implicit* integration scheme, in which the displacement at step $i + 1$ is determined on the basis of the state of the system at step $i + 1$ (Equation 5.49). As can be seen from the solution strategy set out above, an implicit time-stepping solution for a non-linear system is somewhat more complex to solve than an explicit one because it requires iteration at each time step. Nevertheless, implicit schemes have significant benefits, which make them preferable in many circumstances. Most importantly, they are *unconditionally stable*, meaning that there is no limit on time step beyond which errors in the solution are likely to increase without bound. Of course, one would still wish to use a reasonably short time step to ensure that the response is captured with sufficient accuracy.

5.3.5 Extension to MDOF systems

For an MDOF system, if we again assume for simplicity's sake that the non-linearity is restricted to the stiffness terms, our governing equations of motion can be written as

$$m\ddot{x} + c\dot{x} + R(x) = F(t) \tag{5.50}$$

where x is the time-varying nodal displacement vector, m is the mass matrix, c is the damping matrix, R is a vector of stiffness forces that vary non-linearly with x and F is the vector of time-varying, externally applied forces.

The step-by-step integration approaches discussed previously can be applied to MDOF systems in a very similar way. The response history is again divided into short time increments and the response is calculated during each increment using a linearised form of the governing equations, based on structural properties determined at the start of the time step. At the end of the time interval, the structural properties must be updated to reflect the current deformation state of the structure. For example, using the central difference method and following through exactly the same procedure as was used for a non-linear SDOF system, the key time-stepping equation (Equation 5.23) can be written for the MDOF system as

$$\left[\frac{m}{\Delta t^2} + \frac{c}{2\Delta t}\right]x_{i+1} = F_i - R_i + \frac{2m}{\Delta t^2}x_i - \left[\frac{m}{\Delta t^2} - \frac{c}{2\Delta t}\right]x_{i-1} \tag{5.51}$$

However, some issues need careful thought when applying time-stepping methods to MDOF systems.

5.3.5.1 Damping

First, to apply Equation 5.51, we obviously need an explicit formulation of the damping matrix c. In Section 3.5.2, Chapter 3, for linear MDOF systems, we considered some ways of defining the damping. The first of these, modal damping, in which a damping ratio is defined for each mode of the system, is inappropriate here because it is not possible to analyse non-linear systems using the mode superposition approach.

The second, Rayleigh damping, formulates the damping matrix as a linear combination of the mass and stiffness matrices – Equation 3.42:

$$c = a_0 m + a_1 k$$

One of the principal advantages of this approach (for linear systems) is that it ensures that the damping matrix can be diagonalised by the system mode shapes in the same way that the mass and stiffness matrices can. However, for non-linear systems we cannot apply this modal diagonalisation anyway, so this aspect is no longer advantageous. Nevertheless, the Rayleigh damping formulation does give us a quick and simple way of generating a damping matrix and so is often used. For a non-linear system we generally use the initial, elastic stiffness matrix when applying Equation 3.42 (Chapter 3), resulting in a time-invariant damping matrix.

Of course, other formulations of damping matrix could be used. In particular, since we are not using a mode superposition approach, there is no need to choose a damping matrix that is diagonalised by the mode shapes – any set of damping coefficients can be used. For instance, it might be desirable to assign different damping levels to different parts of the structural model, such as the foundations. Increasingly, earthquake resistant design of buildings is making use of purpose-designed damper elements to remove energy from the system, and an explicit damping matrix would allow the distribution of such elements within the structure to be correctly modelled.

5.3.5.2 Stability, accuracy and time step

We considered stability and accuracy, and their relation to choice of time step, for SDOF systems in Section 5.3.3. Similar principles apply for MDOF systems, *viz*:

- The timestep should be sufficiently short to capture both the amplitude and period of the motion with sufficient accuracy – typically requiring 20–40 steps per cycle of the shortest period of interest.
- For a conditionally stable scheme, the time step must not exceed some multiple of the shortest period of the system (the value of the multiplier depending on the particular scheme used, and equalling $1/\pi$ for the CDM).

For an SDOF system with only one natural period, the first of these is likely to govern the choice of time step. For a large MDOF model, however, the situation is quite different. The amplification matrix **A** in Equations 5.38 through 5.42 will now contain terms relating to all of the modes of the MDOF system. In order to ensure that its spectral radius is less than one, it will be necessary for the time step not to exceed some multiple of the *shortest* natural period of the system. As a result, stability will very often be the governing criterion in choosing an appropriate time step.

Suppose, for instance, we wish to model a beam-type structure with a first natural period of 1.0 s, by splitting it into a series of finite elements, giving a model with 10 nodal degrees of freedom. In its linear range, the model will have 10 modes of vibration. To capture the first mode response accurately, we might use a time step of 0.025 s (one-fortieth of the fundamental period). We saw in Chapter 4 that the natural frequencies of a simply supported beam vary with the square of the mode number, so, depending on the exact support conditions, the tenth modal frequency might be 100 times the first, that is, the tenth period will be around 0.01 s. If we wish to analyse the structure using the CDM, then to ensure stability we must use a time step shorter than $0.01/\pi = 0.003$ s. This is about an eighth of the time step of 0.025 s required for accuracy. This is not an extreme example; imagine extending this argument to a model of a complex 3D structure, which might possess hundreds or even thousands of degrees of freedom. The time step required to achieve stability can easily become impractically short.

For this reason, the use of explicit, conditionally stable time-stepping schemes for MDOF dynamics problems is comparatively rare. They are generally reserved for the analysis of extremely fast loading events such as impact, where short time steps are needed anyway in order to capture the extreme rapid loading and response accurately. In other circumstances, implicit, unconditionally stable algorithms are normally preferred.

5.3.6 How many non-linear analyses?

A final point that needs to be made about non-linear time-history analyses concerns the usability of the results. We have already discussed that the accuracy of the analysis will depend on how well the non-linear material characteristics have been modelled, and on the correct implementation of an appropriate numerical method, with careful choice of time step, and so on. However, even if all these aspects have been dealt with properly, the results should still be treated with some caution. This is because non-linear behaviour is an inherently unpredictable process, with the progression of non-linearity likely to be sensitive to quite small changes in the loading.

Consider, for example, the bilinear SDOF structure subjected to the El Centro earthquake in Example 5.4. Suppose we were to repeat the analysis but with the earthquake record reversed in time, so that the strong motion occurs near the end. The resulting input motion has exactly the same frequency content and peak amplitude as previously, the only difference now being that the peaks come in reverse order. The resulting response is shown in Figure 5.23, and Table 5.1 compares the peak displacements and accelerations under the reversed record with those achieved previously. This small change in the input has resulted in changes in the peak response parameters of between about 7% and 15% of the original

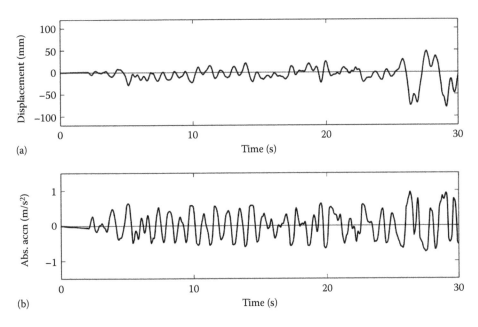

(a)

(b)

Figure 5.23 Response of non-linear SDOF system to El Centro earthquake record reversed in time, determined using the central difference method: (a) relative displacement, (b) absolute acceleration. (Compare with original analysis in Figure 5.20.)

Table 5.1 Peak responses of a non-linear SDOF system to two versions of the El Centro earthquake record

	Peak displacement (mm)		Acceleration (m/s²)	
	Positive	Negative	Positive	Negative
Original signal	57	−74	0.85	−0.81
Reversed signal	48	−79	0.95	−0.74
Percentage change	−15%	+7%	+12%	−9%

values. A somewhat more significant change to the input might be expected to cause even greater variability in the output.

This is quite worrying, especially because the exact form of some future dynamic loading event is unlikely to be known in advance. If a small deviation from the form we assumed can cause a large change in the response, it leaves us in a very uncertain position. To minimise this problem, it will be necessary to perform numerous analyses in which the input loading is varied within some reasonable bounds, and then look at the resulting variability of the peak structural responses. For example, in a seismic design it would be necessary to perform repeat analyses using a suite of earthquake ground motions chosen either to be representative of the ground conditions at the site, or to provide a match to a target spectrum.

The question that then naturally arises is: how many analyses are sufficient? The answer is likely to vary with the circumstances. In seismic design, it is common to use relatively few repeats. For example, the European seismic code, Eurocode 8 (EC8), specifies that the engineer may take either the worst case of three non-linear analyses or the mean of seven. Given the substantial safety factors included in the design process, this is likely to give safe results. However, researchers generally use rather larger suites of load cases to give fuller information on the likely range of responses – if one performs, say, 30 independent analyses, it is possible to determine a robust mean response and also a standard deviation, giving a clear indication of the spread of the data. These can then be used to come up with a *characteristic* value, that is, one with a specified (usually small) probability of exceedence.

5.4 APPROXIMATE TREATMENT OF NON-LINEARITY

The methods of dealing with non-linear dynamics introduced in Section 5.3 are relatively complex. Until quite recently, they were regarded as suitable only for specialists, or as academic research tools. With the increased availability of finite element codes able to perform non-linear analysis efficiently and reliably, and present the results in a digestible way, the use of such methods is increasing. Nevertheless, they still tend not to be the first tool for which the analyst reaches.

There are several reasons for this. Any analysis is only as accurate as the input information, and the non-linear structural properties are likely to be the subject of significant uncertainty. Additionally, the outputs of non-linear analysis can be highly sensitive to the precise sequence of non-linear events occurring during the loading, as we saw in Section 5.3.6. Even if it is accurate and robust, non-linear time-history analysis is in some ways rather a cumbersome tool. It tends to create very large amounts of data, much of which will not be used because the time-stepping analysis generates response data at every step, while designers are generally interested mainly in the peak response.

For these reasons, the idea of using simplified, approximate methods to estimate the influence of non-linearity on dynamic response is attractive. One such method, commonly used in the field of earthquake engineering, is considered here.

5.4.1 Ductility-modified spectra

We saw in Section 2.4.7, Chapter 2 that a dynamic load (particularly an earthquake load) can be conveniently defined by a response spectrum, which summarises the peak response to the loading of linear SDOF systems, as a function of their natural periods and damping ratios. At first, it is hard to see how this could be extended to non-linear systems, since the key element of the method is the elastic natural period of the system. In practice, however, it is possible to come up with an approximate modification of the approach to allow for non-linear response.

It is important to note that we can only do this as part of an integrated design process in which certain rules are followed. The basis of the approach is that, by permitting a structure to suffer some damage short of collapse and so respond in a non-linear way, we can reduce the accelerations (and therefore the forces) for which the structure must be designed. As we will see, the amount of force reduction is related to the *ductility* the structure undergoes, that is, how far it is pushed beyond its elastic limit. While reducing the design loads on the structure is obviously attractive from the point of view of economy, there is a price to be paid; we must ensure the structure is laid out and detailed in such a way as to ensure that the required ductility level can be achieved without catastrophic failure of the structure (seismic codes of practice give guidance on how to achieve this). Additionally, we are accepting that, when the design loadcase happens, the structure will suffer damage and may subsequently have to be demolished and replaced.

The first step in deriving a ductility-modified spectrum is to define exactly what we mean by the ductility of our structure. Consider an EPP SDOF system. Referring to Figure 5.24, the displacement ductility μ can be expressed as

$$\mu = \frac{x_{max}}{x_y} \tag{5.52}$$

Next, we invoke the *equal displacement rule* – based on empirical observations, this rule states that, for structures having a reasonably long elastic period, the peak displacement experienced by a yielding structure is approximately equal to that of a non-yielding structure having the same elastic period.

Now for an elastic SDOF system, we saw in Section 2.4.7, Chapter 2 that the peak, or spectral, acceleration S_a could be related to the maximum displacement S_d because both occur when the mass is at its extreme point, at which instant the velocity is zero. There is therefore no damping force, so equating the inertia and stiffness forces gives

$$mS_a = kS_d \;\rightarrow\; S_a = \frac{k}{m}S_d = \omega_n^2 S_d \tag{5.53}$$

For a yielding structure deformed beyond its elastic limit, the equal displacement rule tells us that the spectral displacement will be the same as in the elastic case, so that the maximum

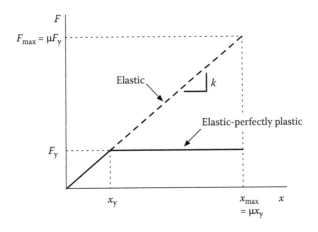

Figure 5.24 Definition of displacement ductility μ for an elastic-perfectly plastic SDOF system.

displacement $x_{max} = S_d$. However, because we are outside the elastic range, this displacement is no longer proportional to the stiffness force. Instead, we have

$$R = kx_y = k\frac{S_d}{\mu} \qquad (5.54)$$

For dynamic equilibrium, this force must still balance the inertia force, so:

$$S_a = \frac{k}{m}\frac{S_d}{\mu} = \frac{1}{\mu}\omega_n^2 S_d \qquad (5.55)$$

Together, the equal displacement rule and Equation 5.55 tell us that, for a longer-period yielding structure, the displacement is similar to that of the equivalent non-yielding structure, but the spectral acceleration, inertia force and stiffness force are all reduced by the ductility factor μ.

This relationship doesn't hold for very short period structures. An idealised, zero-period structure is completely rigid, and so does not deform at all; the mass simply moves with the ground, with no amplification or attenuation. The acceleration it experiences is therefore equal to the ground acceleration, and is the same whether the structure is elastic or EPP.

The question that then arises is how to make the transition between these two cases – what is the shortest period at which the equal displacement rule can be applied, and what happens between this period and the zero period? These questions have been answered in a number of ways. Figure 5.25 presents the approach that is now quite widely adopted, including in the current European seismic code, Eurocode 8. Here, the equal displacement rule is assumed to apply at all periods beyond the left-hand end of the constant acceleration plateau on the spectrum, that is, for periods greater than 0.2 s in the example shown. At a period of 0 s, all the spectra converge on the peak ground acceleration, regardless of the ductility level, and a linear transition is assumed between 0 and 0.2 s.

A note of caution: the spectra shown are not directly compatible with Eurocode 8 – they have been plotted by taking the EC8 type 1 elastic spectrum for ground type A (rock) and scaling by various values of ductility factor. In fact, EC8 uses a slightly different scaling factor, called the *behaviour factor* q; this is really a measure of the force reduction compared to an elastic structure rather than displacement ductility, though Figure 5.24 shows that the two are equivalent for an EPP SDOF system. Additionally, EC8 introduces some additional features such as a further modifying factor on the zero period spectral acceleration, and the imposition of a minimum design acceleration for long-period, high-ductility structures. These complicating factors have been omitted here to maintain the clarity of the presentation.

Use of the ductility-modified spectrum is almost as simple as the elastic spectrum introduced in Chapter 2. For a given elastic period and chosen ductility level, we simply read off the spectral acceleration and use this as the basis for calculating the design force on the structure. The key difference arises if we wish to calculate the peak displacement of the system from the spectral acceleration; we can see from Equation 5.55 that, as well as dividing by the square of the circular natural frequency (as for elastic systems), we must also multiply by the ductility factor.

Example 5.5

Consider SDOF representations of two buildings. Model A, representing a mid-rise structure, has elastic stiffness 8×10^6 N/m and mass 50 t. Model B, representing a stiff, low-rise building, has elastic stiffness 40×10^6 N/m and mass 10 t. Both structures have

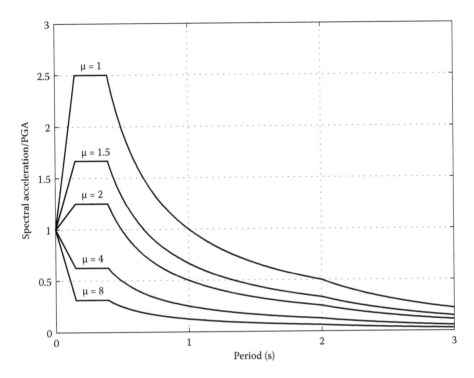

Figure 5.25 Ductility-modified response spectra (sometimes called *design spectra*) for an elastic-perfectly plastic SDOF system. Codes allow engineers to design structures to withstand the reduced acceleration levels so long as they also ensure that the structure can meet the specified ductility level without collapse.

5% damping and are designed as ductile systems that can sustain substantial post-yield deformations without collapse. They are subjected to an earthquake load defined by the ductile spectra introduced earlier, scaled by a peak ground acceleration of 4 m/s² (these are shown in Figure 5.26, scaled and zoomed on the area of interest). Compare their design forces and displacements if they respond elastically, and if they are allowed to sustain a ductility of $\mu = 2$.

First, calculate the elastic natural periods:

$$T_A = 2\pi\sqrt{\frac{m}{k}} = 2\pi\sqrt{\frac{50\times10^3}{8\times10^6}} = 0.50\text{ s}, \quad T_B = 2\pi\sqrt{\frac{m}{k}} = 2\pi\sqrt{\frac{10\times10^3}{40\times10^6}} = 0.10\text{ s}$$

For the elastic case, the spectral accelerations can be read off the top curve ($\mu = 1$) of Figure 5.26, and the forces and displacements calculated using Equation 5.53. By chance, the two periods give the same spectral accelerations, though the forces and displacements are much lower for model B, due its lower mass and higher stiffness.

Model A: Spectral acceleration $S_a = 8.0$ m/s²

Inertia force $= mS_a = 50 \times 8.0 = 400$ kN

Peak displacement $= \dfrac{mS_a}{k} = \dfrac{400\times10^3}{8\times10^6} = 0.050$ m $= 50$ mm

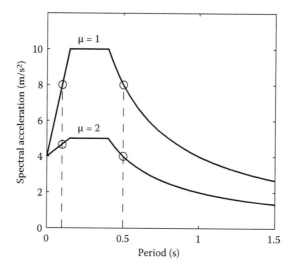

Figure 5.26 Spectra for Example 5.5 (these are a scaled and zoomed version of those in Figure 5.25).

Model B: Spectral acceleration S_a = 8.0 m/s²

Inertia force = mS_a = 10 × 8.0 = 80 kN

$$\text{Peak displacement} = \frac{mS_a}{k} = \frac{80 \times 10^3}{40 \times 10^6} = 0.002 \text{ m} = 2 \text{ mm}$$

Now consider how these results change when a ductility of μ = 2 is permitted. We follow the same procedure as above but now using the lower of the two curves in Figure 5.26 and noting that, in accordance with Equation 5.55 we must scale the calculated displacements by the ductility factor.

Model A: Spectral acceleration S_a = 4.0 m/s²

Inertia force = mS_a = 50 × 4.0 = 200 kN

$$\text{Peak displacement} = \mu\frac{mS_a}{k} = 2 \times \frac{200 \times 10^3}{8 \times 10^6} = 0.050 \text{ m} = 50 \text{ mm}$$

Model B: Spectral acceleration S_a = 4.7 m/s²

Inertia force = mS_a = 10 × 4.7 = 47 kN

$$\text{Peak displacement} = \mu\frac{mS_a}{k} = 2 \times \frac{47 \times 10^3}{40 \times 10^6} = 0.0024 \text{ m} = 2.4 \text{ mm}$$

The effect of ductility can be seen by expressing the ductile system results as percentages of the elastic ones:

	Model A	Model B
Acceleration	50%	59%
Force	50%	59%
Displacement	100%	118%

We see that, for longer-period systems such as model A, accelerations and design forces are reduced in direct proportion to the ductility factor while displacements are unchanged. For very short period systems such as model B, the reduction in design forces is less marked and the ductility causes some increase in peak displacement.

5.4.2 Application to strain hardening and MDOF structures

The theory underlying ductility-modified spectra is sound for SDOF EPP systems. In reality, of course, the structural behaviour is not likely to follow this idealised model. Yielding will often involve the formation of plastic hinges, in which yielding progresses through a cross-section, resulting in a more gradual yield point. In addition, most materials strain-harden with increasing plastic deformation, giving them some positive post-yield stiffness. The more realistic behaviour is compared to the idealised EPP model in Figure 5.27.

While the continuing resistance to deformation after yield can be seen as a good thing, implying greater resistance to collapse, it needs to be treated with care in analysis and design. If design forces have been deduced from a ductility-modified spectrum, then these may well turn out to be smaller than the actual forces generated by the design loading. This can be seen from Figure 5.27; if we assume the equal displacement rule still applies to the strain hardening system, then it will generate a larger force than the EPP one. In other words, a ductility factor of μ on the displacement does not necessarily result in a reduction in forces by the same factor.

In practice, the ductility-modified spectrum approach based on EPP behaviour is nevertheless widely used; the designer must then ensure that the EPP assumption is not violated by too great a margin and/or that the design allows for the possibility of larger forces.

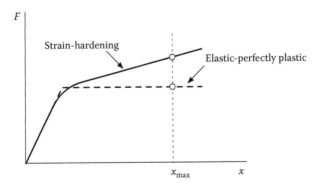

Figure 5.27 Comparison of idealised elastic-perfectly plastic behaviour with more realistic material behaviour, including strain hardening.

We saw in Chapter 3 how the response spectrum approach could be applied to MDOF systems. This approach relied on the principle of linear superposition – splitting the MDOF motion into its component modes of vibration, computing the response of each mode and then combining them to give the total response. Of course, the superposition principle is no longer valid for non-linear systems, calling into question whether such an approach can be used with a ductility-modified spectrum.

A further problem is how to define ductility in an MDOF system. In an SDOF system, there is only one stiffness element that can yield, giving a clear, unambiguous value of yield deformation, but in an MDOF system it is more likely that different elements will yield sequentially, giving a piecewise-linear force-deformation characteristic.

An example is shown in Figure 5.28; as the lateral loads are increased, plastic hinges form in the beams of the multi-storey building in sequence from the bottom to the top. The final loss of stiffness occurs when hinges form at the bases of the columns. Clearly, the level of ductility at the maximum displacement is quite sensitive to how the yield displacement is defined. One possibility would be to take it as the first point at which there is a significant reduction in stiffness (point 1 in Figure 5.28); the disadvantage of this is that the structure still has substantial further load-carrying capacity beyond this point, so we are a long way from the EPP assumption on which the theory is based. A common alternative is to define a best-fit EPP line, as shown dashed in Figure 5.28, and use the single yield point of this line as the basis for ductility calculations. However, this underestimates the initial elastic stiffness of the system and will give very different answers to the previous approach. In design, an engineer is likely to design a structure of this sort with the aim of making all the beam hinges form at a similar value of lateral load, so that points 1–4 on the graph are quite close together and the difference between the actual and the bilinear curves is not too large. However, there is a limit to how well this can be achieved – the difficulties discussed above can be reduced, but not eliminated.

In spite of these concerns, the ductility-modified spectrum approach is quite widely used in the seismic analysis and design of MDOF systems. Experience has shown that it gives acceptable results so long as the engineer is able to design the structure so that yielding and ductility under an extreme load event is reasonably evenly distributed. If this is not the case, then this approach can give unrealistic and unsafe results. In all cases, it needs to be used with care, and with a clear appreciation of its limitations.

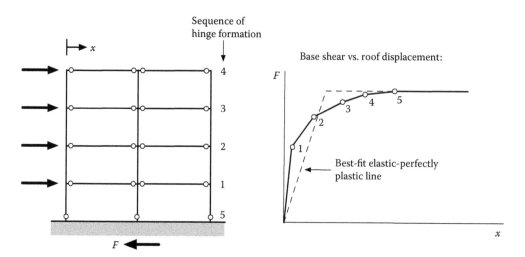

Figure 5.28 Typical yield sequence in an MDOF structure: as the lateral loads are increased in amplitude, hinges form in successive storeys, causing a stepwise reduction in overall lateral stiffness.

KEY POINTS

- Non-linearity in structural behaviour can be caused by variations in the mass, damping or stiffness of the system under consideration, or through the development of such large motions that the usual small-deformation assumptions no longer hold. Of these, reduction in stiffness through yielding is the one normally of most interest when considering dynamic response to extreme loads.
- Methods introduced previously for linear systems cannot be applied to non-linear systems. With variable stiffness, a structure no longer possesses fixed natural frequencies and the principle of modal superposition is not valid for non-linear systems. Frequency domain methods are therefore not applicable.
- Solution generally makes use of direct integration methods, in which the equations of motion are solved at a series of time steps, with the aid of some assumption about how the motion varies between time steps.
- Direct integration methods can be classified as either *explicit* (in which the displacement at a step is expressed entirely in terms of response parameters at previous steps) or *implicit* (in which the displacement at a step is expressed in terms of other response parameters at that step, requiring an iterative solution). Explicit methods have the advantage of simplicity but implicit methods are often preferred due to their superior stability.
- Non-linear time-stepping analysis needs to be done with great care. The time step needs to be chosen to ensure accuracy and (for explicit methods) stability of the solution, and multiple analyses are likely to be necessary to account for the sensitivity of the non-linear system to small variations in the input loads.
- In earthquake engineering, the use of ductility-modified response spectra offers a simple and quick alternative to direct integration methods. However, it is only an approximate method and can give inaccurate answers, especially for MDOF systems in which the non-linearity is not evenly distributed.

TUTORIAL PROBLEMS

Many of the methods presented in this chapter are not amenable to hand calculation and so are not covered in depth in the following problems. To gain a greater understanding, the best approach is to get hold of a user-friendly finite element program and experiment with a *very* simple structural model, trying out different solvers, looking at the effect of varying time step, and so on.

1. Figure P5.1 shows an SDOF mass-spring system restrained by a frictional damper which slips at a force $F_s = 2.0$ kN. A harmonic displacement of amplitude $X = 25$ mm is imposed on the mass. Calculate the energy dissipated in the damper and find the equivalent viscous dashpot coefficient if the loading frequency is: (a) 1 Hz and (b) 5 Hz.

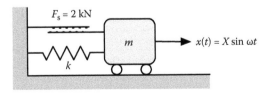

Figure P5.1

If the mass m and stiffness k are related to the load frequency by $\omega^2 = k/m$, what force is required to achieve the imposed displacement?

2. A non-linear viscous damper provides a retarding force related to the velocity across it as described by Equation 5.6. When the velocity is 0.2 m/s the damper force is 400 N, and at 0.55 m/s it is 600 N. Calculate the damper parameters c and α, and specify the units of c. If the damper is subjected to a sinusoidally varying displacement of amplitude 0.1 m and frequency 2.5 Hz, what is the peak damper force?

3. An elastic-perfectly plastic (EPP) spring has initial stiffness 50 kN/m and yield force 5 kN. It is subjected to sinusoidal displacement cycles of amplitude 0.2 m. Sketch the force-displacement hysteresis curve for (a) the first cycle and (b) any subsequent cycle of displacement. In each case, calculate the energy loss due to the plastic deformation.

4. The EPP spring in Problem 3 is replaced by a bilinear one with the same initial stiffness, but a post-yield stiffness of 5 kN/m. It is subjected to the same imposed displacement as in Problem 3. Sketch the new force-displacement hysteresis curves and find the co-ordinates of the first, second and third yield points.

5. A yielding structure is modelled by an elastic-perfectly plastic SDOF system with mass 500 kg and initial stiffness 1 MN/m. Suggest an appropriate time step for an analysis using the central difference method. In an alternative analysis, the structure is modelled as an MDOF system with the same fundamental natural frequency and a highest natural frequency 50 times the fundamental. How does this affect the choice of time step?

6. Two alternative designs are considered for a structure, represented as an SDOF system with mass 4000 kg. In design A, the system has elastic natural period 1.5 s and the structure is detailed so that it can sustain a displacement ductility of 4 without collapse. In design B, the structure is made more stiff but also less ductile, so that its natural period is 0.8 s and its ductility limit is 1.5. Using the design spectra of Figure 5.25 with a peak ground acceleration of 5 m/s², compare the base shear and peak displacements experienced by the two structures.

Chapter 6

Fourier analysis and random vibrations

6.1 INTRODUCTION

In most of the first five chapters of this book, we have considered dynamics problems in which the properties of both the structure being shaken and the excitation are known explicitly, or can be estimated in a deterministic way. We have then used a variety of mathematical techniques to find the dynamic response of the structure to the excitation. In many real-world applications, however, we cannot know the exact form of the loading or the dynamic response. Unpredictable motions of this sort are known as *random vibrations*.

For example, a tall building will be continuously subjected to lateral forces due to wind, which are likely to vary with both time and elevation. Of course, we cannot predict how the wind on the structure will vary on a particular day in the future. However, through measurement and modelling, we can estimate the likely values of many properties of the wind load, for instance the mean wind speed and the fractions of time at which this mean value is exceeded by a certain amount. We can then build a statistical model of the wind load, which, for instance, might take the form of the probability of a certain wind speed being exceeded within a specified time window. In random vibration analysis, we aim to determine the statistical properties of the structural response to a random load from the statistics of the excitation.

A key component of the methods used to do this is the Fourier transform, an extremely powerful mathematical tool that enables us to examine the frequency composition of any dynamic signal. This has some features in common with approaches we have seen before, in which forces or motions are considered in terms of how they are distributed across frequencies, rather than over time. We will consider its application to both deterministic and random signals.

Random vibration analysis is a large topic in its own right. The language and the mathematical techniques are rather different from those we have seen thus far, and it will only be possible to introduce some of the fundamental ideas and methods here. We start with some definitions relating to random processes and probability. We then introduce two of the fundamental tools of signal analysis: the autocorrelation function, which describes how well correlated one part of a time series is with another; and the Fourier transform, which converts a time domain signal into one in terms of frequency. These in turn lead into two ways of defining a dynamic system and analysing its response to a dynamic load: the impulse response function, a time domain representation, and the transfer function, its analogue in the frequency domain. This chapter should enable you to:

- Understand some of the basic terminology of probability and random processes,
- Compute autocorrelation functions of dynamic signals (both deterministic and random),

- Evaluate simple Fourier transforms by hand,
- Understand the use of Fourier transforms in vibration analysis,
- Compute power spectral densities and understand their relationship to autocorrelations,
- Define a linear dynamic system using either an impulse response function or a transfer function,
- Determine the response statistics or a dynamic system from the statistics of a random excitation applied to the system.

6.2 PROPERTIES OF RANDOM VARIABLES AND PROCESSES

Fourier analysis and random vibration theory use a somewhat different vocabulary to the dynamics we have covered thus far. Before we get into too much detail, we need to define some terminology.

6.2.1 Deterministic and random variables

A *deterministic* variable or signal is one that can be defined by some function or algorithm so that, if we know the time t, we can determine the value of the variable $x(t)$. For example, the variable $x(t) = \sin t$ is obviously deterministic. Most of the methods adopted thus far in this book have focused on structures excited by deterministic loads. The exception is the earthquake load case. An individual earthquake time history obviously does not conform to the above definition of a deterministic signal, and we have dealt with this in earlier parts of this book using analysis methods such as Duhamel's integral, which bear some similarity to methods introduced later in this chapter.

A *random* variable owes its form to some underlying random process. While it may be possible to ascertain statistical properties such as the mean of the signal, it is not possible to predict its value at some future time t. Many environmental loads such as wind, wave and earthquake are random in nature, and any experimental vibration measurement will include a random component in the form of electrical noise. Of course, random variables arise in all sorts of other contexts, too, for example, the day-to-day fluctuation in stock market indices.

Whereas the analysis methods presented in previous chapters are mostly suited to deterministic systems, in this chapter we will consider methods that are applicable to both deterministic and random variables.

6.2.2 Discrete and continuous random variables

A *discrete* random variable is one that can take any one of a finite set of values $x_1, x_2, x_3...$ $x_i....$ The value taken on any particular occasion is independent of previous values and cannot be predicted in advance, but it may be possible to determine or estimate the probability of each value occurring. Examples include the usual standard introductory probability cases, such as the value obtained by rolling a die, where each of the six possible outcomes 1, 2, 3, 4, 5, 6 has probability 1/6.

A *continuous* random variable x can take any one of an infinite number of values, either spread over an infinite range or in a band. An example would be the height of a wave in the North Sea. This might typically vary within the range 0 to 25 m; the lower limit is certain (we cannot have a negative wave height), but the upper limit can only be an estimate because a higher value is always possible, even if extremely unlikely.

Because a continuous variable can take any one of an infinite number of possible values, the probability of achieving a specific value is infinitesimally small and cannot be written down. Instead, the best we can do is to express the probability of x lying between two values a and b. For instance, in the wave height example, if we took readings of the maximum wave height in each of 1000 consecutive hours and found that in 100 of those hours the maximum height was between 10 and 12 m, we could estimate the probability of this size of wave to be 0.1.

The distribution of probabilities for a continuous random variable is conveniently expressed using a *probability density function* (PDF) $f(x)$. The PDF is defined such that

$$p(a \leq x \leq b) = \int_a^b f(x)\,dx \tag{6.1}$$

and since the probability of all possible values must be unity, it follows that

$$p(-\infty \leq x \leq \infty) = \int_{-\infty}^{\infty} f(x)\,dx = 1 \tag{6.2}$$

A related function is the *cumulative probability density* $P(x)$, which is the integral of the PDF:

$$P(x) = \int_{-\infty}^{x} f(x)\,dx \quad \text{or} \quad f(x) = \frac{dP(x)}{dx} \tag{6.3}$$

This represents the probability of all values of x up to a certain value, for instance $P(a) = p(x \leq a)$. We can then express the probability of x lying within a certain range in terms of $P(x)$ as follows:

$$p(a \leq x \leq b) = \int_a^b f(x)\,dx = \int_{-\infty}^{b} f(x)\,dx - \int_{-\infty}^{a} f(x)\,dx = P(b) - P(a) \tag{6.4}$$

Figure 6.1 illustrates this relation for an arbitrary PDF.

The most widely used PDF is the normal, or Gaussian, distribution, with its characteristic bell-curve shape. This is defined as

$$f(x) = \frac{1}{\sigma\sqrt{2\pi}} \exp\left(-\frac{(x-\mu)^2}{2\sigma^2}\right) \tag{6.5}$$

where μ and σ are constants. Its exponential form means that it tends to zero as x tends to $\pm\infty$, but never reaches zero. Figure 6.2 shows normal probability density functions for three different values of σ. We will consider the significance of μ and σ more fully in the next section. For now, simply note that the distribution is symmetrical about the peak at $x = \mu$ (taken as zero in this case), and that larger values of σ result in a lower, flatter bell-curve. However, the area under the curve equals unity (as it must for any PDF) whatever the values of μ and σ.

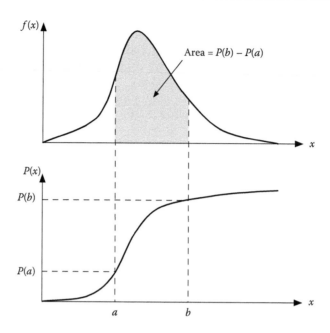

Figure 6.1 Relationship between probability density function *f(x)*, cumulative probability density *P(x)* and the probability that the variable *x* takes a value between *a* and *b*.

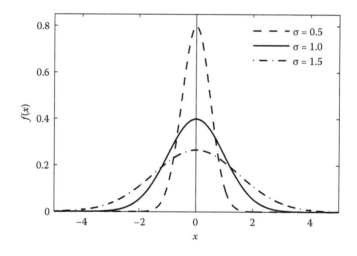

Figure 6.2 Plots of the normal (Gaussian) probability density function for a process with mean $\mu = 0$ and three different values of standard deviation σ.

Much of the power of the normal distribution stems from the *central limit theorem*, which states that if x_k is a random variable formed by summing k other random variables, each with its own PDF (which may have any form), then as k tends to infinity the PDF of x_k converges to a normal distribution. In fact, the convergence is generally quite rapid, so that the approximation to a normal distribution is reasonable even when k is not very large. For this reason, the normal distribution provides a good estimate of the PDF of many natural and physical processes.

6.2.3 Statistical properties of random variables

If we take many (some large number n) samples of a discrete random variable, then we would expect the number of occurrences of the value x_i to be np_i, where $p_i = p(x_i)$. The mean μ of x (in probability theory, also known as the expected value $E(x)$) can then be calculated as the summed values of x divided by the number of samples:

$$\mu = E(x) = \frac{x_1(np_1) + x_2(np_2)\ldots + x_i(np_i)\ldots}{n} = \sum_{i=1}^{n} x_i p_i \tag{6.6}$$

Similarly, for a continuous variable:

$$\mu = \int_{-\infty}^{\infty} x f(x)\,\mathrm{d}x \tag{6.7}$$

In many applications, x can take both positive and negative values with equal probability – as can be seen in the normal distributions of Figure 6.2. In these cases, the mean will be zero, which tells us nothing about the magnitude of the values, and the mean square can be a useful property:

$$E(x^2) = \sum_{i=1}^{n} x_i^2 p_i \quad \text{or} \quad E(x^2) = \int_{-\infty}^{\infty} x^2 f(x)\,\mathrm{d}x \tag{6.8}$$

It is also useful to have a measure of how the results are distributed around the mean – do they all lie very close to the mean value or are they widely spread? This can be quantified by the *variance*, usually denoted by σ^2, defined as the expected value of the squared differences between individual values and the mean:

$$\sigma^2 = \mathrm{Var}(x) = E[(x - \mu)^2] \tag{6.9}$$

Expanding this expression and noting that, since μ takes a fixed value, $E(\mu) = \mu$:

$$\sigma^2 = E(x^2 - 2\mu x + \mu^2) = E(x^2) - 2\mu E(x) + \mu^2 = E(x^2) - \mu^2 \tag{6.10}$$

In words, the variance equals the mean square minus the square of the mean. Finally, the square root of the variance, σ, is known as the *standard deviation*.

Returning to the normal distribution, the significance of the parameters μ and σ is now clear. That μ is the mean of the random variable can be clearly seen in the plots of Figure 6.2. We have already seen that σ is an indicator of the width of the PDF; we now see more specifically that it is a measure of the average distance of x from the mean. For the normal distribution, it is possible to determine the probability of a value of x falling within a chosen number of standard deviations of the mean:

- One standard deviation, that is, x lies in the range $(\mu - \sigma)$ to $(\mu + \sigma)$: $p = 0.683$
- Two standard deviations, that is, x lies in the range $(\mu - 2\sigma)$ to $(\mu + 2\sigma)$: $p = 0.954$
- Three standard deviations, that is, x lies in the range $(\mu - 3\sigma)$ to $(\mu + 3\sigma)$: $p = 0.997$

6.2.4 Two random variables

If we sample two discrete random variables x and y in pairs, then it is possible to define a probability of achieving each possible pair of values, say $p(x_i, y_j)$. Similarly, for a continuous random variable one can define a joint PDF $f(x, y)$ encompassing all possible combinations of x and y. The joint PDF can be visualised as a three-dimensional surface of the form shown in Figure 6.3.

Obviously, for each of x and y we can calculate independently the various statistical properties defined above. We can also define joint properties in a similar way to the single-variable case. The probability of x and y lying between certain values is now the volume beneath the surface between those values. For instance, the probability of x taking a value between a and b while y takes a value between c and d is

$$p[(a \leq x \leq b),(c \leq y \leq d)] = \int_c^d \int_a^b f(x,y)\, dx\, dy \tag{6.11}$$

where the similarity to Equation 6.1 is obvious. If we define some function $g(x, y)$ of the two random variables x and y, then the expected value of g can be defined in a manner analogous to Equation 6.7:

$$E[g(x,y)] = \int_{-\infty}^{\infty} \int_{-\infty}^{\infty} g(x,y).f(x,y)\, dx\, dy \tag{6.12}$$

Often we are interested in assessing the degree of independence between x and y. To do this, we can define the *covariance* of x and y as

$$\sigma_{xy} = E[(x - E(x))(y - E(y))] = E(xy) - E(x)E(y) \tag{6.13}$$

and the *correlation coefficient* as

$$\rho_{xy} = \frac{\sigma_{xy}}{\sigma_x \sigma_y} \tag{6.14}$$

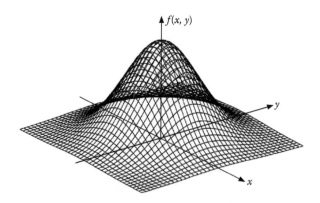

Figure 6.3 A joint probability distribution for two random variables x and y.

If x and y are statistically independent, then $E(xy) = E(x)E(y)$, in which case both the covariance and the correlation coefficient are zero. On the other hand, if there is a direct, deterministic relationship between x and y, then the correlation coefficient takes a value of unity.

Example 6.1

Consider a single-degree-of-freedom (SDOF) system subjected to harmonic base shaking $y = Y\sin \omega t$, resulting in a vibration of the mass $x = X\sin(\omega t + \phi)$, where ϕ is a phase angle. Explore how the correlation between the two signals varies with the phase angle.

In this example, x and y are not truly independent, random signals because both are defined by a functional relationship in terms of the time variable t. We will therefore treat t as our single independent variable. Of course, t can take any positive value, but since the signals are periodic with period $T = 2\pi/\omega$ we can evaluate the necessary averages over any one period.

Suppose we consider the signal values at some arbitrary moment in time, say $t = \tau$, where $0 \leq \tau \leq T$. Since τ can take any value between 0 and T with equal probability, its PDF is simply $f(\tau) = 1/T$. Now find the expected value of the product xy:

$$E[x(\tau)y(\tau)] = \int_0^T x(\tau)y(\tau).f(\tau)\,d\tau = \int_0^T X\sin(\omega\tau + \phi).Y\sin\omega\tau \frac{1}{T}d\tau$$

$$= \frac{XY}{T}\int_0^T \sin\omega\tau.(\sin\omega\tau\cos\phi + \cos\omega\tau\sin\phi)\,d\tau = \frac{XY}{2}\cos\phi$$

By inspection, a sine wave has expected (mean) value zero so $E(x) = E(y) = 0$ and, using Equation 6.13, the above result is also the covariance of x and y. We can also calculate the variance of x alone and y alone (Equation 6.10):

$$\sigma_x^2 = E(x^2) = \int_0^T (X\sin(\omega\tau + \phi))^2 \frac{1}{T}d\tau = \frac{X^2}{2}, \quad \sigma_y^2 = \frac{Y^2}{2}$$

Finally, the correlation coefficient can be found from Equation 6.14:

$$\rho_{xy} = \frac{\sigma_{xy}}{\sigma_x\sigma_y} = \frac{(XY\cos\phi)/2}{(X/\sqrt{2}).(Y/\sqrt{2})} = \cos\phi$$

The correlation coefficient thus varies smoothly with the phase angle. When the phase angle is zero, we have $\rho_{xy} = 1$ and the signals are perfectly correlated; this is to be expected as they are sine waves having the same period and no phase difference. When $\phi = 180°$, we get $\rho_{xy} = -1$, a condition which we again describe as perfect correlation; in this case, the signals are in anti-phase. When $\phi = 90°$ or $270°$ we have $\rho_{xy} = 0$ and the two signals are uncorrelated.

6.2.5 Properties of random processes

The term *random process* is generally taken to describe an ensemble of dependent random variables that may be functions of one or more independent variables – in structural engineering applications these are often time and position. For example, suppose we measured

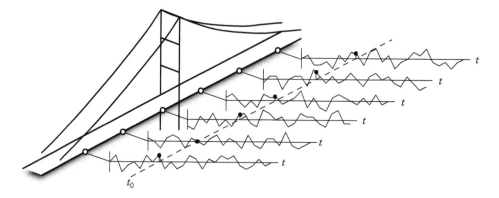

Figure 6.4 An ensemble of wind speed measurements along the deck of a suspension bridge. A time average can be computed for each measurement position; alternatively, an ensemble average can be obtained by taking the mean across all positions at a chosen time, say t_0.

wind speed at numerous (say, n) points along the span of a long suspension bridge, as illustrated in Figure 6.4. The measurements are likely to include a significant steady component, but also a randomly fluctuating one due to air turbulence. The random part will vary with time t but also with span position. We thus have an ensemble of n time-dependent random variables $x_i(t)$.

For a process such as this, we can calculate two kinds of average. The *ensemble average* is found by averaging across the ensemble of signals at a particular instant in time, say t_0. Here we are effectively calculating the mean, or expected value $E(x)$, of a set of discrete random variables:

$$E(x) = \frac{1}{n} \sum_{i=1}^{n} x_i(t_0) \tag{6.15}$$

Alternatively, we can find the *time average* x by choosing a single signal, say x_k, and averaging over its duration:

$$\bar{x} = \lim_{T \to \infty} \frac{1}{T} \int_{0}^{T} x_k(t)\, dt \tag{6.16}$$

A random process is called *stationary* if its ensemble average $E(x)$ is independent of time, that is, the same value is obtained regardless of the choice of t_0. Of course, for a process to be truly stationary, this definition implies it carries on over an infinite time span. This is rarely the case for engineering processes but we often assume that a process is stationary over its duration.

A process for which the time and ensemble averages are equal is known as *ergodic*. Knowing that a process is ergodic is useful because it means we can assume that one sample is completely representative of the ensemble that makes up the random process. An ergodic process must also be stationary – the ensemble average can only be equal to the time average if it is also unchanging over time. However, the reverse is not always true – a stationary process is not necessarily ergodic.

6.3 AUTOCORRELATION

One of the fundamental mathematical tools for the analysis of random signals, the autocorrelation function provides a measure of the correlation between different parts of a time-varying signal. Its evaluation makes use of a mathematical technique called convolution, which will be introduced first.

6.3.1 Convolution integral

We define the convolution between two time domain functions $x(t)$ and $y(t)$ as

$$x(t) * y(t) = \int_{-\infty}^{\infty} x(t) y(t + \tau) \, dt \qquad (6.17)$$

In words, the convolution is the integral of the product of two signals, one shifted in time relative to the other, as a function of the time shift τ. In fact, we have already seen an example of a convolution integral – Duhamel's integral (Section 2.4.6, Chapter 2); autocorrelation will provide us with another. However, before we move onto that, it is useful to try to visualise the process of convolution, which is best done through an example.

Example 6.2

Find the convolution of the functions:

$$x(t) = X \quad \text{when } 0 < t < a$$

$$y(t) = Y \quad \text{when } 0 < t < b$$

where $a > b$, and the functions are zero outside these ranges. The two pulses are plotted in Figure 6.5a.

Following the definition in Equation 6.17, we must multiply the original function x by a time-shifted version of the function y. The relation between the two functions is shown for various values of the time shift in Figure 6.5b. The potentially confusing part here is which direction the y pulse moves as τ increases. However, this can be quite easily resolved if we note that y is non-zero only when its argument $(t + \tau)$ lies between 0 and b, that is, when t is between $-\tau$ and $(b - \tau)$. So, when τ is negative, the y pulse is shifted to the right of its original position, and as τ is increased, it slides leftward, as indicated by the arrows in the diagrams. For any value of τ, the convolution integral is then the area of the product formed by multiplying the pulses where they overlap – shown shaded.

$\tau < -a$:	No overlap – $x*y = 0$
$-a < \tau < b - a$:	$x*y = XY(a + \tau)$
$b - a < \tau < 0$:	$x*y = XYb$
$0 < \tau < b$:	$x*y = XY(b - \tau)$
$\tau > b$:	No overlap – $x*y = 0$

The convolution integral therefore takes the form shown in Figure 6.5c. Its largest values occur at time shifts where the two pulses interact most strongly, and it drops to zero when there is no interaction between the signals.

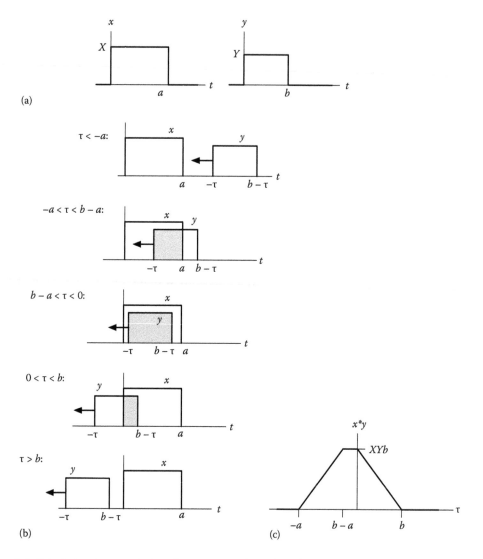

Figure 6.5 Calculating the convolution of two rectangular pulses: (a) original pulses, (b) visualising the process for different values of the shift variable τ and (c) the resulting convolution function.

6.3.2 Autocorrelation function

Now consider a stationary process x. We define the *autocorrelation function* R_{xx} as the covariance of the signal and a time-shifted version of itself. Since the process is stationary, R_{xx} is independent of the absolute time t and so is a function only of the separation time. Denoting this time-shift by τ:

$$R_{xx}(\tau) = E(x(t)x(t + \tau)) \tag{6.18}$$

If the process is ergodic, then this ensemble average could also be written in the form of a time average of a single signal x_k:

$$R_{xx}(\tau) = \lim_{T \to \infty} \frac{1}{T} \int_{-T/2}^{T/2} x_k(t) x_k(t+\tau) \, dt \qquad (6.19)$$

Comparing this definition to Equation 6.17, we can see that the autocorrelation is in fact the convolution of $x(t)$ with itself. In a similar way to Example 6.2, we can visualise the evaluation of the autocorrelation as the sliding of one version of the signal past another by varying the time shift τ. Figure 6.6 shows an arbitrary function $x(t)$ and a time-shifted version; the integrand in the above equation can be plotted by multiplying point-by-point. The value of R_{xx} for this particular value of time shift is then the mean of the resulting signal. To form the full autocorrelation function, we need to repeat this exercise for each possible value of the time shift τ.

As its name implies, the autocorrelation function provides a measure of the extent to which a signal value at some point in time is correlated to its value at other times. With little or no temporal correlation, we would expect R_{xx} to reduce in value quickly as the magnitude of the shift τ increases; for a strongly correlated signal, on the other hand, R_{xx} will remain non-zero at large τ.

A few points about R_{xx} can be noted from the above equations. It is always an even function, that is, imposing a negative time shift will give the same result as a positive shift of equal magnitude. When $\tau = 0$, the signal is simply multiplied by itself without any time shift and the autocorrelation is then equal to the mean square, or *average power* of the signal.

The term *power* can be confusing here; it is not being used in the accepted engineering sense of work per unit time and does not have units of Watts – its units are the square of

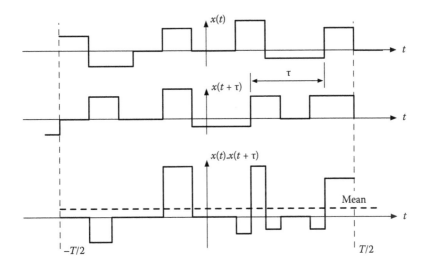

Figure 6.6 A typical step in the calculation of autocorrelation: the original signal (top graph) is shifted by a time τ (middle graph). The two are then multiplied together (bottom graph) and the mean of this time series is the value of autocorrelation for the chosen value of τ.

those of the underlying signal. In signal analysis, power is generally understood to refer to the square of a quantity, and to provide a measure that is related to the energy of the system. (Note that many energy expressions do involve a square: strain energy is proportional to extension squared, kinetic energy to velocity squared.)

Lastly, referring to Equation 6.10, if the process has zero mean then the average power is also equal to the variance of the signal.

Example 6.3

Find and plot the autocorrelation function of the square wave signal of amplitude A and period $2T$ shown in Figure 6.7a.

Figure 6.7b shows the signal shifted by an amount τ such that $0 \le \tau \le T$. Since the signal repeats regularly, the autocorrelation evaluated over any one period will be the same as that evaluated over all time. Within one period, the integral of Equation 6.19 can be split into four consecutive segments:

$$R_{xx}(\tau) = \frac{1}{2T}\left(\int_0^{T-\tau} A^2\,dt + \int_{T-\tau}^{T} (-A^2)\,dt + \int_{T}^{2T-\tau} A^2\,dt + \int_{2T-\tau}^{2T} (-A^2)\,dt \right)$$

$$= \frac{1}{2T}\left(A^2(T-\tau) + (-A^2)\tau + A^2(T-\tau) + (-A^2)\tau \right) = A^2\left(1 - \frac{2\tau}{T} \right)$$

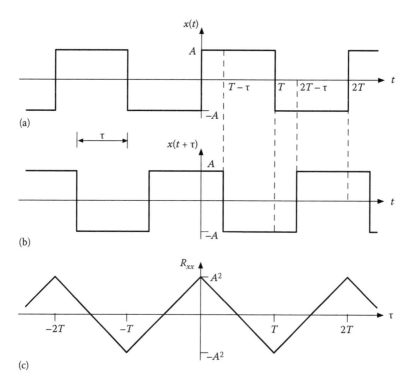

Figure 6.7 Computing the autocorrelation of a square wave: (a) original waveform, (b) time-shifted version and (c) autocorrelation function.

By a similar process, when $T \leq \tau \leq 2T$ we get:

$$R_{xx}(\tau) = A^2 \left(\frac{2\tau}{T} - 3 \right)$$

When $\tau = 2T$ the shifted signal comes back into perfect alignment with the original one and as τ is increased further, the above calculations repeat themselves, giving an autocorrelation that is periodic. It should also be obvious that the autocorrelation at negative values of τ follows an identical pattern.

The result is plotted in Figure 6.7c. As would be expected, the average power is A^2 and, since the signal has zero mean, the variance also equals A^2. Note that for any periodic signal, the autocorrelation *must* also be periodic because time shifts that are multiples of the period will always bring the two signals back into perfect alignment.

Example 6.4

Suppose now that, instead of changing sign after every interval T as in the previous example, we have a random signal where, after every interval T, there is an equal chance that the signal takes the value $+A$ or $-A$ (Figure 6.8a). Find the autocorrelation function.

Since the signal now contains a probabilistic component, we will use the expected value definition (Equation 6.18) to evaluate the autocorrelation. Consider separately the two possibilities that the time shift τ may be less than or greater than the interval T.

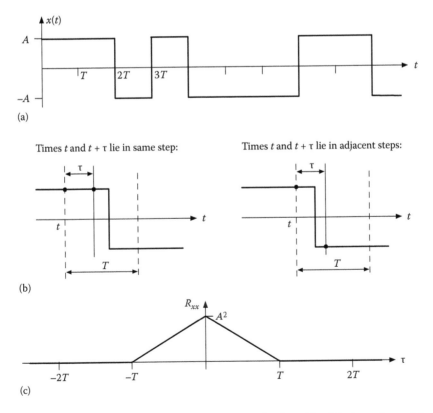

Figure 6.8 Computing the autocorrelation of a random train of pulses: (a) original waveform, (b) comparing the time shift to the pulse interval for the case $\tau < T$ and (c) autocorrelation function.

If $\tau < T$ then, as illustrated in Figure 6.8b, the two times t and $t + \tau$ may either fall within the same interval or in adjacent intervals. Since our starting point t can fall anywhere within the interval with equal probability, we can express the probabilities that the second point lies on a new step, or on the same step as the first, as

$$p(\text{new step}) = \frac{\tau}{T}, \quad p(\text{same step}) = 1 - \frac{\tau}{T}$$

Next we need to evaluate the probabilities of the four possible pairs of values at times t and $t + \tau$: (A, A), $(A, -A)$, $(-A, A)$ and $(-A, -A)$. Consider first the probability that the signal takes the value A both at time t and at $t + \tau$. This can be written as

$$p(A, A) = p(A \text{ at time } t).p(\text{same step at time } t + \tau)$$

$$+ p(A \text{ at time } t).p(\text{new step at time } t + \tau).p(\text{new step has value } A)$$

$$= \frac{1}{2} \cdot \left(1 - \frac{\tau}{T} \right) + \frac{1}{2} \cdot \frac{\tau}{T} \cdot \frac{1}{2} = \frac{1}{2} - \frac{1}{4} \frac{\tau}{T}$$

The same result applies for $p(-A, -A)$. For the two cases where the signal has changed value over the time shift, this can only be achieved if the shift has caused us to move to the next step, so

$$p(A, -A) = p(A \text{ at time } t).p(\text{new step at time } t + \tau).p(\text{new step has value } -A)$$

$$= \frac{1}{2} \cdot \frac{\tau}{T} \cdot \frac{1}{2} = \frac{1}{4} \frac{\tau}{T}$$

and again the same result is achieved for $p(-A, A)$. Reassuringly, it can be seen that the probabilities of the four combinations sum to 1. We are now ready to calculate the autocorrelation. Since the signals vary stepwise rather than continuously, we can do this as a discrete summation rather than an integral.

$$R_{xx}(\tau) = E(x(t)x(t + \tau)) = \sum x(t)x(t + \tau).p[x(t)x(t + \tau)]$$

$$= (A.A).p(A, A) + (-A. - A).p(-A, -A) + (A. -A).p(A, -A) + (-A.A).p(-A, A)$$

$$= 2A^2 \left(\frac{1}{2} - \frac{1}{4} \frac{\tau}{T} \right) - 2A^2 \frac{1}{4} \frac{\tau}{T} = A^2 \left(1 - \frac{\tau}{T} \right)$$

Now consider the case where $\tau > T$. In this case we can be certain that the times t and $t + \tau$ lie within different steps, that is,

$$p(\text{new step}) = 1, \quad p(\text{same step}) = 0$$

All four possible pairs of values now have equal probability and, following through the same calculation procedure as before yields the result $R_{xx}(\tau) = 0$.

Remembering that it is always an even function (i.e. it takes the same values for positive and negative time shifts), the autocorrelation function takes the form in Figure 6.8c. We see that the autocorrelation at $\tau = 0$ (that is, the average power of the signal) is the same as for the square wave in the previous example, but that the random nature of the signal results in a rapid reduction away from $\tau = 0$. For time shifts greater than the step length, the signal is completely uncorrelated ($R_{xx} = 0$). This demonstrates the earlier statement that a random signal has correlation only over short time scales. Indeed, if we consider a case where the switching between random values takes place continuously, that is, $T \to 0$, then the triangle in Figure 6.8c reduces to a sharp spike at the origin and the signal is completely uncorrelated over any non-zero time offset.

It is interesting to note that, in Example 6.4, we were able to determine an explicit auto-correlation function even though the signal from which it was derived was defined probabilistically, so that its precise form could not be known. This gives us a way into estimating the response of structures to random dynamic loads, described in probabilistic terms; we will see in subsequent sections that we can obtain frequency domain descriptions of random dynamic signals from their autocorrelations, and can use these to estimate the dynamic response of systems.

6.4 FOURIER ANALYSIS

Fourier analysis is the fundamental tool that allows us to take time-varying vibration signals of any form and examine them in terms of their frequency content. The key to the method is the Fourier transform, an extension of the Fourier series idea introduced in Chapter 1.

6.4.1 Fourier transform

We saw in Section 1.5.2 (Chapter 1) and Appendix C that a periodic function with period T can be written as a series of sine and/or cosine terms known as a Fourier series, which breaks the signal down into a set of harmonic components at a base frequency $\omega_0 = 2\pi/T$ and at integer multiples of it

$$x(t) = \frac{A_0}{2} + \sum_{n=1}^{\infty} \left(A_n \cos \frac{2\pi nt}{T} + B_n \sin \frac{2\pi nt}{T} \right) \tag{6.20}$$

where

$$A_n = \frac{2}{T} \int_{-T/2}^{T/2} x(t) \cos \frac{2\pi nt}{T} \, dt, \quad B_n = \frac{2}{T} \int_{-T/2}^{T/2} x(t) \sin \frac{2\pi nt}{T} \, dt \tag{6.21}$$

The coefficients A_n and B_n tell us the size of each frequency component. Therefore, if we plot the coefficients against the frequencies of the sinusoids with which they are associated, we get a sort of spectrum, showing us how the power of the signal is distributed across frequencies (Figure 6.9a). The frequency spacing of the harmonics is

$$\Delta\omega = \frac{2\pi}{T} \tag{6.22}$$

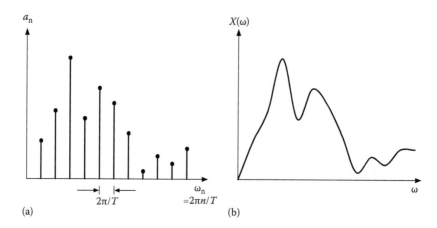

Figure 6.9 Representing a signal in the frequency domain: (a) Fourier series coefficients for a periodic signal and (b) Fourier transform of a non-periodic signal.

Clearly if the period of our signal becomes large, the frequency spacing of the spectral lines in Figure 6.9a becomes very small. In the limit, if we let $T \rightarrow \infty$, then two things happen: first, the frequency spacing becomes infinitesimally small, so that we achieve a continuous distribution rather than a set of discrete points (Figure 6.9b), and secondly the signal ceases to be periodic because if one period has infinite duration then it is never repeated. This result is expressed mathematically as the *Fourier transform*, which gives the frequency distribution $X(\omega)$ corresponding to a time-domain signal $x(t)$:

$$X(\omega) = \int_{-\infty}^{\infty} x(t)e^{-i\omega t} \, dt \tag{6.23}$$

This can be thought of as the replacement of Equation 6.21 for non-periodic signals. Similarly, we can replace the discrete sum in Equation 6.20 by the inverse Fourier transform, or Fourier integral:

$$x(t) = \frac{1}{2\pi} \int_{-\infty}^{\infty} X(\omega)e^{i\omega t} \, d\omega \tag{6.24}$$

Equations 6.23 and 6.24 are often known as the *Fourier transform pair*, allowing switching of a function between its time and frequency domain representations. Often you will see Fourier transform pairs written in the form:

$$x(t) \Leftrightarrow X(\omega)$$

The positioning of the factor of $1/2\pi$ is arbitrary. Some texts place it in the first equation of the transform pair and some in the second; others insert a factor of $1/\sqrt{(2\pi)}$ in both equations. The choice made here is probably the most usual. The derivation of Equations 6.23 and 6.24 is presented a little more fully in Appendix C, along with a table of Fourier transforms of some common functions.

Example 6.5

Find the Fourier transform of the decaying exponential function $x = e^{-at}$ $(t \geq 0)$, shown in Figure 6.10a.

Using Equation 6.23:

$$X(\omega) = \int_{-\infty}^{\infty} x(t) e^{-i\omega t} \, dt = \int_{0}^{\infty} e^{-at} e^{-i\omega t} \, dt = \left[\frac{-1}{a+i\omega} e^{-(a+i\omega)t} \right]_{0}^{\infty} = \frac{1}{a+i\omega}$$

The result is thus a complex quantity, which is better understood in terms of magnitude and phase. Multiplying both numerator and denominator by $a - i\omega$, we can write

$$X(\omega) = \frac{a - i\omega}{a^2 + \omega^2}$$

Recall from Section 1.3.2 (Chapter 1) that we can represent a harmonic motion at a frequency ω as a vector in the complex plane, rotating at angular speed ω, as shown in Figure 6.10b. Simple trigonometry then gives the magnitude and phase angle as

$$|X| = \frac{1}{\sqrt{a^2 + \omega^2}}, \quad \tan\phi = -\frac{\omega}{a}$$

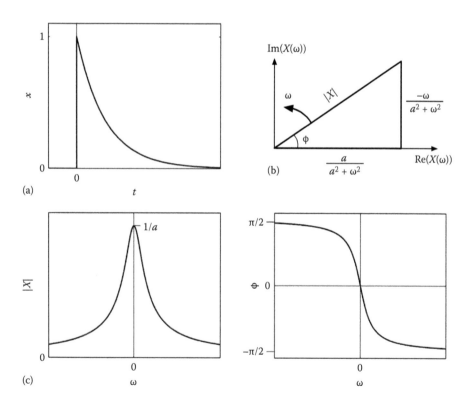

Figure 6.10 Calculation of Fourier transform of a negative exponential: (a) the time domain function, (b) converting the complex result to a magnitude and a phase angle and (c) the resulting Fourier transform.

These results are plotted in Figure 6.10c. In some instances, we will mainly be concerned with the magnitude plot because this gives us the key information about how the amplitude of the signal is distributed across frequencies. However, the phase plot also contains important information. If we wished to reconstruct the time domain signal from the Fourier transform, we would need to know not just the amplitude of the signal at various frequencies, but also how these different components are aligned relative to each other along the time axis – this, of course, is the information contained in the phase part of the transform.

One thing to note about the Fourier transform is that it exists for both positive and negative values of frequency. The idea of a negative frequency might seem odd, but it becomes less so if we recall that $\cos(-\omega t) = \cos(\omega t)$ and $\sin(-\omega t) = -\sin(\omega t)$. The components at negative frequencies can therefore be easily translated into equivalent positive-frequency ones.

6.4.2 Properties of Fourier transforms

When manipulating Fourier transforms, it is helpful to be able to call on the following properties, which are quoted here without proof; standard proofs are available in many maths textbooks. In the following, a is a constant, t_0 is a fixed time and ω_0 a fixed frequency.

Linearity: If $x(t) \Leftrightarrow X(\omega)$ then $ax(t) \Leftrightarrow aX(\omega)$.

Scaling: If $x(t) \Leftrightarrow X(\omega)$ then $x(at) \Leftrightarrow \dfrac{1}{|a|}X\left(\dfrac{\omega}{a}\right)$.

Time shift: If $x(t) \Leftrightarrow X(\omega)$ then $x(t - a) \Leftrightarrow e^{-i\omega a}X(\omega)$.

Frequency shift: If $x(t) \Leftrightarrow X(\omega)$ then $x(t)e^{i\omega_0 t} \Leftrightarrow X(\omega - \omega_0)$.

Differentiation: If $x(t) \Leftrightarrow X(\omega)$ then $\dot{x}(t) \Leftrightarrow i\omega X(\omega)$.

Convolution: If $x(t) \Leftrightarrow X(\omega)$ and $y(t) \Leftrightarrow Y(\omega)$ then $x(t)^*y(t) \Leftrightarrow X(\omega)Y(\omega)$.

Modulation: If $x(t) \Leftrightarrow X(\omega)$ and $y(t) \Leftrightarrow Y(\omega)$ then $x(t)y(t) \Leftrightarrow \dfrac{1}{2\pi}X(\omega) * Y(\omega)$.

6.4.3 The Dirac delta function

To apply Fourier transforms to vibration problems we need to use the Dirac delta function, sometimes also called the unit impulse, denoted by $\delta(t)$. This function, which was briefly introduced in Section 2.4.5 (Chapter 2), takes the value zero for any non-zero value of t, and at $t = 0$ it can be thought of as a short pulse of infinite height and infinitesimal width, such that it encloses a unit area (Figure 6.11a). In mathematical terms, this can be expressed as

$$\delta(t) = 0 \quad t \neq 0 \tag{6.25}$$

and

$$\int_{-\infty}^{\infty} \delta(t)\,dt = 1 \tag{6.26}$$

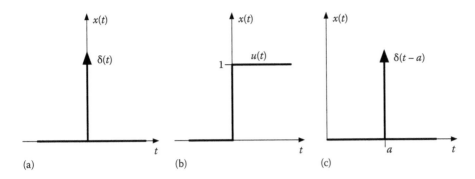

Figure 6.11 (a) The Dirac delta function, (b) the Heaviside unit step function and (c) the delta function at time $t = a$.

The delta function is related to another useful discontinuous function, the Heaviside or unit step function, $u(t)$ (Figure 6.11b). This takes the value zero when $t < 0$ and unity when $t > 0$:

$$u(t) = \begin{cases} 0 & t < 0 \\ 1 & t > 0 \end{cases} \tag{6.27}$$

The step function has zero gradient except at $t = 0$ and so we easily see that

$$\delta(t) = \frac{du(t)}{dt} \tag{6.28}$$

It is easy to shift the delta function to a position other than $t = 0$; the function $\delta(t - a)$ is zero except at $t = a$, as shown in Figure 6.11c. Furthermore, if we consider the integral of the product of the delta function with some other function of t:

$$\int_{-\infty}^{\infty} x(t).\delta(t - a)\,dt = x(a) \tag{6.29}$$

So, this operation has the effect of picking out the value of the continuous function $x(t)$ at the single point $t = a$, a procedure known as *sifting*.

Now consider the relationship of the delta function to the Fourier transform. Using the definition of the Fourier transform (Equation 6.23) together with the sifting property of the delta function (Equation 6.29), we can write the transform of the delta function as

$$X(\omega) = \int_{-\infty}^{\infty} \delta(t)e^{-i\omega t}\,dt = e^{-i\omega.0} = 1 \tag{6.30}$$

This simple result tells us that a very short impulse in time is equivalent to a very large number of frequency components of equal magnitude, since the Fourier transform takes the same value whatever the frequency.

Next, we evaluate the inverse Fourier transform of a delta function in the frequency domain, say at frequency ω_0. Applying Equation 6.24:

$$x(t) = \frac{1}{2\pi} \int\limits_{-\infty}^{\infty} \delta(\omega - \omega_0) e^{i\omega t} \, d\omega = \frac{1}{2\pi} e^{i\omega_0 t}$$

We therefore have the Fourier transform pair:

$$e^{i\omega_0 t} \Leftrightarrow 2\pi\delta(\omega - \omega_0) \tag{6.31}$$

Recalling the relationships between the complex exponential and sine and cosine functions (Equations 1.13 and 1.14), we can now deduce the following Fourier transform pairs:

$$\cos\omega_0 t = \frac{1}{2}\left(e^{i\omega_0 t} + e^{-i\omega_0 t}\right) \Leftrightarrow \pi\left[\delta(\omega - \omega_0) + \delta(\omega + \omega_0)\right] \tag{6.32}$$

$$\sin\omega_0 t = -\frac{i}{2}\left(e^{i\omega_0 t} - e^{-i\omega_0 t}\right) \Leftrightarrow i\pi\left[-\delta(\omega - \omega_0) + \delta(\omega + \omega_0)\right] \tag{6.33}$$

We thus see that a vibration signal that varies sinusoidally with time can be represented in the frequency domain as a pair of sharp spikes (delta functions), one at the frequency of the sinusoid and another at minus that frequency. It follows that a signal whose Fourier transform comprises a series of such spikes is the superposition of several sinusoids at different frequencies. Reconstruction of the time domain signal from the Fourier transform requires both the amplitudes of the different components and the phase terms, which tell us the time lags between them.

6.4.4 Practical use of Fourier transforms

The Fourier transform provides an extremely powerful way of understanding and analysing vibration data. For instance, it is frequently used in experimental modal analysis to determine the frequency of vibrations from a time domain signal collected by a displacement transducer or an accelerometer.

Whereas we have determined Fourier transforms analytically from continuous time-domain functions, of course real data are recorded by sampling a continuous signal at a series of discrete points equally spaced in time. The integrals of Equations 6.23 and 6.24 must therefore be replaced by a summation, a formulation known as the discrete Fourier transform (DFT). In fact, the computation is normally performed using a very clever algorithm called the fast Fourier transform (FFT), which can reduce the number of calculations by several orders of magnitude.

As an illustrative example of the use of the FFT, consider the signal shown in Figure 6.12a. In this case, the signal shows a displacement, measured in mm, but it could equally well be any measurement quantity, such as acceleration in m/s². The signal is the sum of two harmonic terms, one at a frequency of 2 Hz and the second at 5 Hz, having half the amplitude of the first. The signal has been distorted by some random measurement noise (artificially generated in this case). It is just about possible to detect the 2 Hz component by visual inspection, for instance by counting the number of zero crossings in the 10-s time window. However, the noise makes this difficult to do accurately, and renders it near impossible to make out the 5 Hz component.

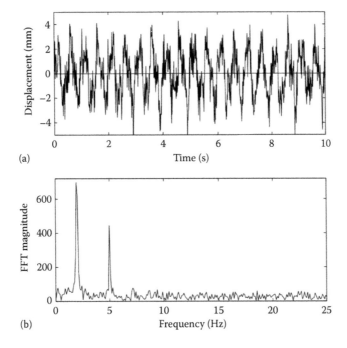

Figure 6.12 (a) A noisy time-domain vibration signal and (b) its Fourier transform.

In Figure 6.12b, the signal is displayed in the frequency domain as calculated by the FFT. Only the magnitude (the modulus of the real and complex parts) is shown here, and not the phase. This now clearly shows the presence of two dominant frequencies, and that the 2 Hz one is the larger. However, the ratio of 2 between the magnitude of the two harmonic terms is not accurately captured, with the strength of the 2 Hz peak appearing to have been split across two adjacent frequency ordinates. We can also see that the noise is evenly distributed across a large range of frequencies. (See Section 6.5.1 for a discussion of white noise.) Note also that the numerical values of FFT magnitude are not particularly helpful, bearing little relation to the magnitudes of the time domain signal.

A proper treatment of discrete methods is beyond this text. Various complications are introduced, such as the interaction between sampling rate and frequency resolution, and the effect of finite duration sampling on the frequency domain representation. See the reading list for guidance on where to learn more about these issues.

6.5 POWER SPECTRAL DENSITY

We have seen that the Fourier transform of many common vibration signals is a complex quantity. This should not alarm us – we know that the imaginary part represents the phase of a particular frequency component, that is, how far it needs to be shifted along the time axis when plotting it in the time domain. However, often we are more interested in the magnitude of a dynamic quantity than its phase, and complex algebra can be fiddly to deal with. A useful quantity, therefore, is the *power spectral density* (PSD). For a signal $x(t)$, the power spectral density $S_{xx}(\omega)$ is defined as

$$S_{xx}(\omega) = X(\omega).X^*(\omega) \tag{6.34}$$

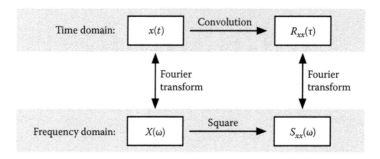

Figure 6.13 Relationship of a time-domain signal to its Fourier transform, autocorrelation and power spectral density.

Here $X^*(\omega)$ is the complex conjugate of $X(\omega)$, the Fourier transform of $x(t)$. So if $X(\omega)$ has the general form $a + ib$, then $X^*(\omega) = a - ib$. It is then clear that the PSD must always be a real quantity since

$$X(\omega).X^*(\omega) = (a + ib).(a - ib) = a^2 - i^2 b^2 = a^2 + b^2 = \left|X(\omega)\right|^2$$

Recalling that the autocorrelation function $R_{xx}(\tau)$ is the convolution of $x(t)$ with itself, and applying the convolution property of Fourier transforms stated in Section 6.4.2, we see that

$$x(t) * x(t) \Leftrightarrow [X(\omega)]^2$$
$$R_{xx}(\tau) \Leftrightarrow S_{xx}(\omega)$$

(6.35)

Therefore, the autocorrelation and the PSD of a signal are a Fourier transform pair. These relationships are summarised in Figure 6.13.

Example 6.6

We return to the random signal introduced in Example 6.4/Figure 6.8a in which, after every time interval T, there is an equal chance that the signal takes the value $+A$ or $-A$. Find and plot its PSD.

Since the signal is probabilistic in form, there is no straightforward way of computing its Fourier transform, making the application of Equation 6.34 problematic. However, we have already determined its autocorrelation in Example 6.4, so we can find the PSD from this, using Equation 6.35. The autocorrelation function is a triangular pulse (Figure 6.14a). The transform of a triangular pulse of unit amplitude between times $\tau = \pm a$ is given in Table C.2 of Appendix C as

$$X(\omega) = a \frac{\sin^2(\omega a/2)}{(\omega a/2)^2}$$

The signal in Figure 6.14a has amplitude A^2 rather than unity and pulse time dimension T rather than a. The power spectral density can therefore be written as

$$S_{xx}(\omega) = A^2 T \frac{\sin^2(\omega T/2)}{(\omega T/2)^2}$$

The result is plotted in Figure 6.14b. Note that the PSD is a real, even function.

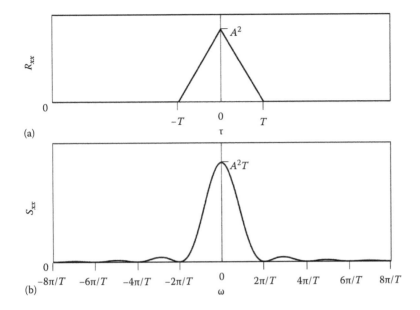

Figure 6.14 (a) Autocorrelation and (b) power spectral density of a random train of pulses.

Now it follows from Equation 6.35 that the autocorrelation can be written as the inverse Fourier transform of the PSD:

$$R_{xx}(\tau) = \frac{1}{2\pi} \int\limits_{-\infty}^{\infty} S_{xx}(\omega)e^{i\omega\tau} \, d\omega$$

As we saw earlier, the value of the autocorrelation when the time shift $\tau = 0$ is equal to the average power (mean square) of the signal. Therefore:

$$P_{ave} = R_{xx}(0) = \frac{1}{2\pi} \int\limits_{-\infty}^{\infty} S_{xx}(\omega)e^{i\omega.0} \, d\omega = \frac{1}{2\pi} \int\limits_{-\infty}^{\infty} S_{xx}(\omega) \, d\omega \qquad (6.36)$$

Therefore, the average power of a signal can be determined either as the zero value of the autocorrelation or from the area under the PSD curve. Equation 6.36 also helps us to clarify the dimensions of the various quantities. The autocorrelation is a measure of power (defined as the square of the signal amplitude) and so has units of signal squared. The PSD is a function of frequency whose integral is a power, and so has units of signal squared divided by frequency. For example, if the signal $x(t)$ is a displacement measured in m, then the autocorrelation has units of m² and the PSD has units of m²/(rad/s).

Note the presence of the $1/2\pi$ factor in Equation 6.36. One way to get rid of it is to note that the frequency f of a vibration is related to ω by $f = \omega/2\pi$ and hence $df = d\omega/2\pi$, allowing us to write Equation 6.36 as

$$P_{ave} = \int\limits_{-\infty}^{\infty} S_{xx}(\omega)\frac{d\omega}{2\pi} = \int\limits_{-\infty}^{\infty} S_{xx}(2\pi f) \, df \qquad (6.37)$$

Lastly, as we saw in Example 6.6, unlike the Fourier transform, the PSD is always both real and even. It is therefore quite common to plot it only for the positive frequency range, giving a single-sided PSD, denoted $W_{xx}(\omega)$. In doing this, in order to preserve the relationship between average power and area under the PSD curve, we scale the PSD by a factor of two, that is, $W_{xx}(\omega) = 2S_{xx}(\omega)$.

Example 6.7

In Example 6.6, we determined a PSD in terms of ω. Write the equivalent PSD in terms of frequency f, giving both the double- and single-sided forms, and show that all three forms give the same average power.

Recall the previous result:

(a) $\displaystyle S_{xx}(\omega) = A^2 T \frac{\sin^2(\omega T/2)}{(\omega T/2)^2} \qquad -\infty < \omega < \infty$

Making the substitution $\omega = 2\pi f$:

(b) $\displaystyle S_{xx}(2\pi f) = A^2 T \frac{\sin^2(\pi f T)}{(\pi f T)^2} \qquad -\infty < f < \infty$

The single-sided spectrum is

(c) $\displaystyle W_{xx}(2\pi f) = 2A^2 T \frac{\sin^2(\pi f T)}{(\pi f T)^2} \qquad 0 \le f < \infty$

Evaluating the required integrals is rather complex, and only the results are quoted here. Taking each of the above forms of the PSD in turn, we get:

(a) $\displaystyle P_{ave} = A^2 T \frac{1}{2\pi} \int_{-\infty}^{\infty} \frac{\sin^2(\omega T/2)}{(\omega T/2)^2}\, d\omega = A^2 T \frac{1}{2\pi} \cdot \frac{2\pi}{T} = A^2$

(b) $\displaystyle P_{ave} = A^2 T \int_{-\infty}^{\infty} \frac{\sin^2(\pi f T)}{(\pi f T)^2}\, df = A^2 T \frac{1}{T} = A^2$

(c) $\displaystyle P_{ave} = 2A^2 T \int_{0}^{\infty} \frac{\sin^2(\pi f T)}{(\pi f T)^2}\, df = 2A^2 T \frac{1}{2T} = A^2$

and, as expected, the result is in each case equal to $R_{xx}(0)$.

6.5.1 White noise

A particularly important form of random signal is known as *white noise*. By definition, this is a random process whose autocorrelation is a delta function at time zero. The electrical noise that normally appears when collecting experimental data from instrumentation is often a close approximation to white noise. An example of a white noise signal is shown in Figure 6.15.

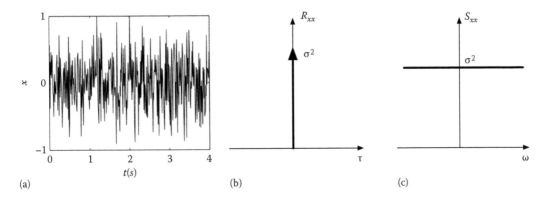

Figure 6.15 Zero-mean white noise: (a) time domain representation, (b) autocorrelation and (c) power spectral density.

To understand the form of the autocorrelation, recall the definition (Equation 6.19):

$$R_{xx}(\tau) = \lim_{T\to\infty} \frac{1}{T} \int_{-T/2}^{T/2} x(t)x(t+\tau)\,dt$$

Now white noise is a zero-mean random process that is equally likely to take a large or small, positive or negative value at any instant. Therefore, when we multiply it by a time-shifted version of itself, the product $x(t)x(t+\tau)$ at any time t is again equally likely to be large or small, positive or negative; the resulting integral must therefore be zero. The only exception is, of course, for the case where the time shift τ is zero, when the signal is simply multiplied by itself. The resulting product is then entirely positive and the integral is equal to the mean square of the signal. We saw in Equation 6.10 that, for a signal with zero mean, the mean square is equal to the variance σ^2, so we can write:

$$R_{xx}(\tau) = \sigma^2.\delta(\tau) \tag{6.38}$$

Now consider the PSD of white noise. We recall Equation 6.30, in which we showed that the Fourier transform of a delta function is unity. Therefore, taking the transform of the autocorrelation gives:

$$S_{xx}(\omega) = \sigma^2 \tag{6.39}$$

Therefore, a genuine white noise signal is one whose power is uniformly distributed across all frequencies.

Figure 6.16 shows the PSD of the noisy vibration signal introduced in Figure 6.12. As we would expect, the PSD shows peaks at the same frequencies as the Fourier transform magnitude plot of Figure 6.12b, but because it is a plot of power density (i.e. signal squared), the peaks stand out more clearly. As we would expect from the argument above, the random noise contributes a small component to the PSD whose amplitude is roughly constant with frequency.

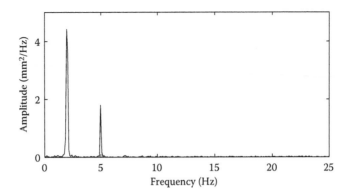

Figure 6.16 Power spectral density of the noisy time-domain vibration signal of Figure 6.12a.

6.6 IMPULSE RESPONSE AND TRANSFER FUNCTION

For any dynamic system, but particularly for those subject to random inputs, it is convenient to be able to come up with a single function that defines its dynamic properties. This enables us to translate directly between the inputs to the system (the loads or other excitations) and the outputs (the structural response), and lends itself readily to block diagram representation. We will see that we can do this either in the time domain, using the impulse response function, or in the frequency domain, using the transfer function. Figure 6.17 summarises the relationships between the various functions. We briefly introduced this approach in Section 2.5, where we used the Laplace transform to express the transfer function in terms of the Laplace variable s. We now use the Fourier transform to express it in terms of frequency ω, a more physically meaningful quantity.

6.6.1 Impulse response function

We have seen in Section 2.4.5 (Chapter 2) that one of the simplest response calculations we can do for a system is to find its response to a short impulse at time $t = 0$. Define the unit impulse function as

$$I(t) = I_0 \delta(t) \tag{6.40}$$

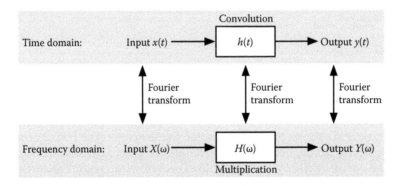

Figure 6.17 Impulse response and transfer function for a linear dynamic system.

where the scalar I_0 has unit value and is there to get the dimensions of the input right, since the delta function is non-dimensional. For instance, if the input to our system is a force then we would expect the unit impulse to have units of force × time, so would set $I_0 = 1$ Ns. The response of a dynamic system to this impulsive load is then termed the *unit impulse response*, $h(t)$.

Next, consider the same dynamic system subjected to a time-varying load. Whatever form the load takes, one way of analysing it is to break it up into a series of short, impulsive loads (Figure 6.18). The kth impulse at time τ_k, with amplitude $I_k = x_k \Delta\tau$, can be written as

$$I_k \delta(t - \tau_k) = x_k \Delta\tau \, \delta(t - \tau_k)$$

and because the system is linear the response to this impulse is scaled accordingly:

$$y_k(t) = x_k \Delta\tau . h(t - \tau_k)$$

Then the response to the full load is the sum of the responses to the constituent impulses:

$$y(t) = \sum_k y_k(t) = \sum_k x_k \Delta\tau . h(t - \tau_k)$$

If we let the interval $\Delta\tau$ tend to the infinitesimal quantity $d\tau$, the summation becomes an integral:

$$y(t) = \int_0^t x(\tau) h(t - \tau) \, d\tau \tag{6.41}$$

Equation 6.41 tells us that the output of a linear dynamic system is the convolution of the input with the system's unit impulse response function, and can be written more compactly as

$$y(t) = h(t)^*x(t) \tag{6.42}$$

It is important to note here that $h(t)$ is the only term related to the system that is required in order to calculate its dynamic response. We can therefore say that the impulse response

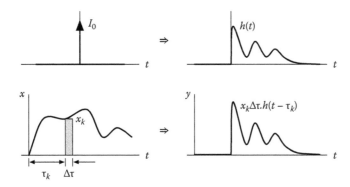

Figure 6.18 Calculating the response y to a dynamic input x using the unit impulse response $h(t)$.

function is a system property that entirely defines the dynamic characteristics of the system under analysis.

(**Note:** Strictly speaking, the convolution integral should have limits of $\pm\infty$. However, if the load function x is defined as starting at time zero, then the integral over negative time is zero. Similarly, if the impulse response function h is zero when its argument is negative, then the integral is also zero for $\tau > t$. So long as these conditions are met, Equation 6.41 is equivalent to a true convolution integral.)

Example 6.8

Find the impulse response function of a linear SDOF system with mass m, damping coefficient c and stiffness k. The input to the system is a time-varying force and the output is a displacement.

To calculate the impulse response, we apply a unit impulsive force at time $t = 0$. The governing equation for the impulse displacement response $h(t)$ is then:

$$m\ddot{h}(t) + c\dot{h}(t) + kh(t) = 1.\delta(t)$$

We have seen before that it is helpful to introduce the non-dimensional parameters

$$\omega_n = \sqrt{\frac{k}{m}}, \quad \xi = \frac{c}{2\sqrt{km}}$$

so that the governing equation becomes

$$m\left[\ddot{h}(t) + 2\xi\omega_n\dot{h}(t) + \omega_n^2 h(t)\right] = \delta(t)$$

We saw in Section 2.4.5 (Chapter 2) that a simple way to solve this equation is to treat it as a free vibration in which the effect of the impulse is simply to impose an initial velocity on the system equal to the magnitude of the impulse divided by the mass (using impulse = change in momentum). We therefore need only solve

$$\ddot{h}(t) + 2\xi\omega_n\dot{h}(t) + \omega_n^2 h(t) = 0 \quad \text{subject to} \quad h(0) = 0, \ \dot{h}(0) = 1/m$$

The general solution to this differential equation is

$$h(t) = e^{-\xi\omega_n t}(A\sin\omega_d t + B\cos\omega_d t), \quad \omega_d = \omega_n\sqrt{1-\xi^2}$$

Applying the zero displacement initial condition gives $B = 0$ and then differentiating:

$$\dot{h}(t) = Ae^{-\xi\omega_n t}(-\xi\omega_n\sin\omega_d t + \omega_d\cos\omega_d t)$$

Then, substituting the initial velocity condition gives $A = 1/m\omega_d$ and the impulse response can be written as

$$h(t) = \frac{1}{m\omega_d}e^{-\xi\omega_n t}\sin\omega_d t$$

6.6.2 Transfer function

If we take the Fourier transform of Equation 6.42, recalling the convolution property from Section 6.4.2, we get

$$Y(\omega) = H(\omega)X(\omega) \tag{6.43}$$

where $H(\omega)$ is the *transfer function* of the system, defined as the ratio of the output to the input in the frequency domain, and sometimes also called the *frequency response function*. The link between transfer function and impulse response is shown schematically in Figure 6.17. Note that, whereas the impulse response function defines the system behaviour in terms of the time domain reaction to a short stimulus, the transfer function gives the steady state response to a harmonic input. Despite these different definitions, both are equivalent ways of characterising the dynamic properties of a system, and they are related through the Fourier transform.

While Equation 6.43 looks very simple, note that the transfer function in many cases may be complex. As we have seen before, the complex form can be interpreted as giving us information about amplitude and phase in a single expression. If the transfer function has the general form

$$H(\omega) = A(\omega) - iB(\omega)$$

then the amplitude and phase are

$$|H(\omega)|^2 = [A(\omega)]^2 + [B(\omega)]^2, \quad \tan\phi = \frac{B(\omega)}{A(\omega)}$$

where a positive value of ϕ implies that the output lags behind the input. Now recall that the PSD of a time series $x(t)$ or $y(t)$ is the square of its Fourier transform. Therefore, squaring each term in Equation 6.43 gives

$$S_{yy}(\omega) = |H(\omega)|^2 S_{xx}(\omega) \tag{6.44}$$

Example 6.9

Find the transfer function of a linear SDOF system with mass m, damping coefficient c and stiffness k when subjected to an external force.

As in Example 6.8, the governing equation for the system can be written either in terms of m, c and k or using the damping ratio ξ and natural frequency ω_n:

$$m\left[\ddot{x}(t) + 2\xi\omega_n\dot{x}(t) + \omega_n^2 x(t)\right] = F(t)$$

To find the transfer function, assume a general harmonic input of the form

$$F(t) = F_0 e^{i\omega t}$$

We would then expect a harmonic output of the form

$$x(t) = H(\omega)F_0 e^{i\omega t}$$

Here only the exponential term is a function of time, so differentiating and substituting into the equation of motion gives

$$m\left[-\omega^2 + 2\xi\omega_n.i\omega + \omega_n^2\right]H(\omega)F_0e^{i\omega t} = F_0e^{i\omega t}$$

from which we deduce that

$$H(\omega) = \frac{1}{m\left[\left(\omega_n^2 - \omega^2\right) + 2i\xi\omega_n\omega\right]}$$

To interpret it as an amplitude and a phase, we wish to express the transfer function in the form $A - iB$. We do this by multiplying the numerator and denominator by the complex conjugate of the term in square brackets to give

$$H(\omega) = \frac{\left(\omega_n^2 - \omega^2\right) - 2i\xi\omega_n\omega}{m\left[\left(\omega_n^2 - \omega^2\right)^2 + (2\xi\omega_n\omega)^2\right]}$$

The amplitude and phase are then

$$\left|H(\omega)\right| = \frac{1}{m\sqrt{\left(\omega_n^2 - \omega^2\right)^2 + (2\xi\omega_n\omega)^2}}, \quad \tan\phi = \frac{2\xi\omega_n\omega}{\left(\omega_n^2 - \omega^2\right)}$$

6.6.3 Application to random vibration analysis

Finally, we will look at how to determine the statistics of the response of a linear system to a random excitation. Consider a system with input $x(t)$, output $y(t)$ and transfer function $H(\omega)$. The input $x(t)$ is stationary and random with mean μ_x and autocorrelation $R_{xx}(\tau)$, both of which are assumed to be known. We can proceed as follows:

1. First we can calculate additional properties of $x(t)$. We saw in Section 6.3.2 that the mean square of a signal is given by the value of the autocorrelation function with a zero time shift, $R_{xx}(0)$. We can then use Equation 6.10 to find the variance:

$$\sigma_x^2 = R_{xx}(0) - \mu_x^2$$

and taking the Fourier transform of $R_{xx}(\tau)$ gives the PSD of the input, $S_{xx}(\omega)$.

2. We can now find the PSD of the system response, $S_{yy}(\omega)$, by multiplying the input PSD by $\left|H(\omega)\right|^2$ (Equation 6.44).

3. The autocorrelation of the response, $R_{yy}(\tau)$, can be found by taking the inverse Fourier transform of $S_{yy}(\omega)$, and the mean square value of the response is given by $R_{yy}(0)$.

4. We can now calculate the mean and variance of the response using:

$$\mu_y = H(0).\mu_x$$

$$\sigma_y^2 = R_{yy}(0) - \mu_y^2$$

It is interesting to note here that, although both the input and the output are random variables, the system is deterministic; the power spectra are uniquely defined, and there is a unique relationship between them.

Example 6.10

An SDOF system has mass $m = 500$ kg, natural frequency $\omega_n = 20$ rad/s and damping ratio $\xi = 0.05$. It is subjected to a wind load that is treated as a zero-mean white noise with variance $A = 10^6$ N^2. Find the power spectral density, mean and variance of the displacement of the SDOF system.

First, consider the loading. With zero mean, the mean square of the loading is simply equal to its variance and, from Equations 6.38 and 6.39, the autocorrelation and PSD are

$$R_{xx}(\tau) = A.\delta(\tau), \quad S_{xx}(\omega) = A$$

The transfer function for a damped SDOF system was found in Example 6.9. Then, applying Equation 6.44 gives the PSD of the response:

$$S_{yy}(\omega) = |H(\omega)|^2 S_{xx}(\omega) = A \left| \frac{1}{m\left[\left(\omega_n^2 - \omega^2\right) + 2i\xi\omega_n\omega\right]} \right|^2 = \frac{A}{m^2\left[\left(\omega_n^2 - \omega^2\right)^2 + (2\xi\omega_n\omega)^2\right]}$$

The autocorrelation of the response can be found by taking the inverse Fourier transform of the PSD:

$$R_{yy}(\tau) = \frac{1}{2\pi} \int_{-\infty}^{\infty} A \left| \frac{1}{m\left[\left(\omega_n^2 - \omega^2\right) + 2i\xi\omega_n\omega\right]} \right|^2 e^{i\omega\tau} d\omega$$

This is not an easy integral to solve. However, since we are trying to find the mean and variance, we do not need to find the autocorrelation for all values of τ. Instead, we first find the mean of the response:

$$\mu_y = H(0).\mu_x = \frac{1}{m\omega_n^2}.0 = 0$$

With zero mean, it then follows that the variance is equal to the mean square of the response, which is given by the autocorrelation evaluated at $\tau = 0$:

$$\sigma_y^2 = R_{yy}(0) = \frac{1}{2\pi} \int_{-\infty}^{\infty} A \left| \frac{1}{m\left[\left(\omega_n^2 - \omega^2\right) + 2i\xi\omega_n\omega\right]} \right|^2 d\omega$$

This is still not a simple integral to solve, but the result can be found in numerous maths textbooks and is

$$\sigma_y^2 = \frac{A}{4m^2\xi\omega_n^3}$$

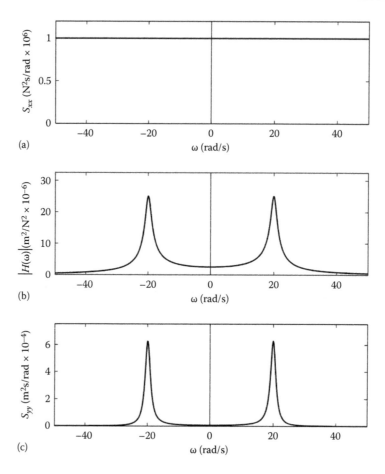

Figure 6.19 Response of an SDOF system to white noise excitation: (a) power spectral density of input force, (b) system transfer function and (c) power spectral density of displacement response.

Inserting numerical values, the expressions for the PSD and variance become:

$$S_{yy} = \frac{1}{(400-\omega^2)^2 + 4\omega^2}\,(\text{m}^2\text{s/rad}), \quad \sigma_y^2 = 0.0125\ \text{m}^2$$

In Figure 6.19, the PSDs of the input and response are plotted, together with modulus of the system transfer function. These take the form one would expect, with clear resonant peaks at the system natural frequency. For a zero-mean process, the variance of the response is equal to the area under its PSD curve, divided by the factor 2π.

It will be noted that even this simple example requires some quite difficult mathematics. For more complex cases, it will be necessary to resort to numerical methods using discrete forms of the autocorrelation and Fourier transform.

6.6.4 Modal testing

The techniques introduced in this chapter are widely used in *modal testing*, in which the dynamic properties of structures are determined by field monitoring. As dynamics has grown in importance, so has the topic of modal testing. This is because analytical predictions of structural properties can be unreliable due to uncertainties over matters such as joint and foundation stiffness, added mass and the effect of non-structural components on dynamic performance. Experimental determination of natural frequencies and damping levels therefore offers far greater certainty than numerical modelling. In addition, modal testing has become far more accessible with the availability of affordable instrumentation and high-quality data analysis software.

In a typical modal testing procedure, the structure is excited dynamically by a measured force provided by a device such as an electrodynamic shaker, which is attached to the structure. The force imparted must be sufficiently large to mobilise measurable motion of the structure, and is normally applied over a range of frequencies. Motion of the structure is most easily measured by accelerometers positioned at a grid of locations across the structure. Using discrete Fourier transforms, the time domain signals can be converted to functions of frequency. The transfer function, which defines the dynamic properties of the system, can then be determined as the ratio of the measured output to the input using Equation 6.43 or 6.44. Finally, dynamic parameters such as natural frequency and damping ratio can be derived by taking an appropriate analytical transfer function such as the SDOF one derived in Example 6.9 and choosing the parameters to give a best fit of the analytical function to the experimentally derived one.

In some cases, for example, very large suspension bridges, the structure may vibrate due to dynamic loads that we cannot control and cannot measure (wind, traffic, etc). In these cases, the above procedure cannot be applied, but it has been shown that very good estimates of dynamic properties can be achieved by assuming the excitation can be approximated as white noise, as was done in Example 6.10. See the Further Reading at the end of this book for more information on modal testing.

KEY POINTS

- A random load or vibration is one that cannot by defined by a mathematical function or algorithm, so that it is impossible to determine its value at some future time. Many environmental dynamic loads are random in nature.
- For random vibrations, it is generally necessary to work in terms of the statistical properties of the key variables, such as mean and variance, rather than their precise values at particular times.
- The autocorrelation of a time history provides a measure of how much correlation there is between signal values at different points in time. It is found by integrating the product of a signal with a time-shifted version of itself. Autocorrelation is always an even function; positive and negative time shifts produce the same result.
- The Fourier transform enables us to express a time-varying signal as a function of frequency. It is normally a complex function that can be interpreted as an amplitude, showing how the magnitude of the signal is distributed across frequencies, and a phase, describing the time lags between the different frequency components.
- The power spectral density (PSD) describes how the power (signal squared) of a vibration record is distributed across frequencies. It is equal to the product of the Fourier transform with its complex conjugate; this results in a real quantity, and thus the phase information is lost. The PSD can also be found by taking the Fourier transform of the signal's autocorrelation function, an approach that is particularly advantageous for random signals.

- A dynamic system such as a mass-spring-damper can be defined in the time domain by an impulse response function, or in the frequency domain by a transfer function. The two definitions are related by the Fourier transform.
- The easiest way to find the response of a dynamic system to a random load is to use the system transfer function to find the PSD of the response from that of the input. Using Fourier transforms, the PSDs can then be converted to their corresponding auto-correlation functions, allowing statistical properties such as mean and variance to be determined.

TUTORIAL PROBLEMS

1. Find the autocorrelation of: (a) $x(t) = \sin \omega t$, (b) the periodic waveform shown in Figure P6.1.

2. Find the mean, mean square and variance of the two waveforms in Problem 1.

3. Find and sketch the Fourier transforms of the three functions plotted in Figure P6.3.

4. A continuous vibration $x(t)$ has Fourier transform $X(\omega)$. A finite-duration signal $y(t)$ is created by recording a segment of $x(t)$ from time zero to time T. Using the modulation property of Fourier transforms (Section 6.4.2) explain how the Fourier transform of $y(t)$ is related to that of $x(t)$. How is this relationship affected by the choice of the sampling duration T?

Figure P6.1

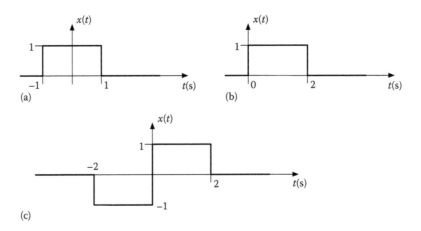

Figure P6.3

5. An undamped SDOF system has mass 2000 kg and natural frequency 10 rad/s. Using the impulse response function method, compute the maximum displacement response to the loading function shown in Figure P6.5, and the amplitude of free vibrations after the loading has ceased.

6. Figure P6.6 shows an SDOF system subjected to a time-varying ground motion $x_g(t)$. Determine the transfer functions relating the following output and input quantities:

   (a) Output = displacement of mass *relative* to ground, input = ground acceleration,

   (b) Output = displacement of mass *relative* to ground, input = ground displacement,

   (c) Output = *absolute* displacement of mass, input = ground displacement.

7. Figure P6.7 shows a massless trolley, connected to a moving support by a spring and a damper, and to a rigid support by a spring. Find the transfer function between the displacement of the trolley and the ground displacement.

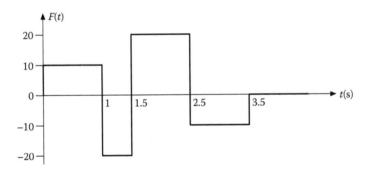

Figure P6.5

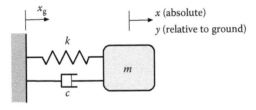

Figure P6.6

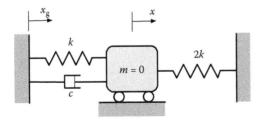

Figure P6.7

8. A mass m is restrained by a viscous damper with coefficient c, but no spring. Find the transfer function between a force applied to the mass, and its displacement, expressing the answer as an amplitude and a phase.

9. A zero-mean, stationary, random force has the autocorrelation function:

$$R_{xx}(\tau) = F_0^2 e^{-|\tau|}$$

Find the average power of the force signal, and its power spectral density.

10. The random force in Problem 9 is applied to an undamped SDOF structure with mass m and natural frequency ω_n. Find the power spectral density of the displacement response of the structure. If $F_0 = 10$ kN, $m = 2000$ kg and $f_n = 2.5$ Hz, sketch the PSD and calculate its amplitude at zero frequency.

11. A random ground vibration is defined in terms of the ground acceleration. It takes the form of white noise with zero mean and variance of 12.0 m²/s⁴. It is applied to an SDOF system with natural frequency $f_n = 4.0$ Hz and damping ratio $\xi = 0.05$. Find the mean and variance of the displacement of the mass relative to the ground.
    You may find the following result useful – for a function of the form

$$H(\omega) = \frac{d}{a + i\omega b - \omega^2 c}$$

the integral of its squared modulus is

$$\int_{-\infty}^{\infty} |H(\omega)|^2 \, d\omega = \frac{\pi d^2}{ab}$$

Background mathematics

Appendix A: Basic trigonometry and the complex exponential

Many dynamic loads and/or motions are harmonic; that is, they can be expressed as sine or cosine functions of time. It is worth understanding the equivalence of various ways of writing this. First, we can define trigonometric functions with reference to a right-angled triangle with sides of length A, B and C (Figure A.1).

$$\sin\theta = \frac{B}{C}, \quad \cos\theta = \frac{A}{C}, \quad \tan\theta = \frac{B}{A} \tag{A.1}$$

Pythagoras's theorem states:

$$A^2 + B^2 = C^2 \tag{A.2}$$

It follows that:

$$\sin^2\theta + \cos^2\theta = \frac{B^2 + A^2}{C^2} = 1, \quad \tan\theta = \frac{\sin\theta}{\cos\theta} \tag{A.3}$$

We also recall the trigonometric identity for the sine of the sum of two angles:

$$\sin(\theta_1 + \theta_2) = \sin\theta_1 \cos\theta_2 + \cos\theta_1 \sin\theta_2 \tag{A.4}$$

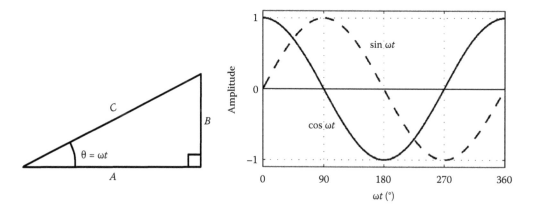

Figure A.1 Definition and plots of sin ωt and cos ωt.

A harmonic motion can be expressed as the sum of a sine and cosine term:

$$x = A \sin \omega t + B \cos \omega t \tag{A.5}$$

This can be manipulated into a form similar to the right-hand side of Equation A.4:

$$x = C\left(\frac{A}{C}\sin \omega t + \frac{B}{C}\cos \omega t\right) = C(\cos\phi \sin \omega t + \sin\phi \cos \omega t)$$

from which we can see that

$$x = C \sin (\omega t + \phi) \tag{A.6}$$

is exactly equivalent to Equation A.5 so long as

$$C = \sqrt{A^2 + B^2} \quad \text{and} \quad \tan\phi = \frac{B}{A} \tag{A.7}$$

Here C is the amplitude of x and ϕ is its phase – a measure of how far it is shifted along the time axis. Note that a cosine wave is simply a sine wave shifted by a phase angle $\phi = \pi/2$ rads (90°) as can be seen from the plots in Figure A.1, or by putting $A = 0$ in Equations A.5 through A.7.

Harmonic motion can also be expressed using the complex exponential. Euler's equation states

$$e^{i\theta} = \cos\theta + i \sin\theta \tag{A.8}$$

Noting that $\cos(-\theta) = \cos\theta$ and $\sin(-\theta) = -\sin\theta$, we can also write

$$e^{-i\theta} = \cos\theta - i \sin\theta \tag{A.9}$$

Rearranging Equations A.8 and A.9 gives the following expressions:

$$\cos\theta = \text{Re}\{e^{i\theta}\} = \frac{1}{2}(e^{i\theta} + e^{-i\theta}) \tag{A.10}$$

$$\sin\theta = \text{Im}\{e^{i\theta}\} = \frac{1}{2i}(e^{i\theta} - e^{-i\theta}) \tag{A.11}$$

where Re and Im refer, respectively, to the real and imaginary parts of the complex function. Using Equation A.11, a motion at angular frequency ω and phase angle ϕ, as in Equation A.6, can be written as

$$x = \text{Im}\{Ce^{i(\omega t + \phi)}\} = \text{Im}\{Ce^{i\omega t}e^{i\phi}\} \tag{A.12}$$

Appendix B: Second order ordinary differential equations

Many dynamic systems are governed by second-order ordinary differential equations (ODEs) of the form:

$$a\frac{d^2x}{dt^2} + b\frac{dx}{dt} + cx = F(t), \quad \text{or} \quad a\ddot{x} + b\dot{x} + cx = F(t) \tag{B.1}$$

where a, b and c are constant coefficients and each dot denotes one differentiation with respect to t. The full solution is the sum of two parts: a *complementary function* obtained by putting the right-hand side equal to zero, and a *particular integral* related to the function $F(t)$.

To obtain the complementary function, we adopt a trial solution of the form

$$x = Ae^{st} \tag{B.2}$$

Substituting into Equation B.1 with the right-hand side set to zero:

$$Ae^{st}(as^2 + bs + c) = 0 \tag{B.3}$$

This is known as the *auxiliary equation*. For a non-trivial solution, the quadratic term inside the brackets must equal zero, giving

$$s = \frac{-b \pm \sqrt{b^2 - 4ac}}{2a} \tag{B.4}$$

The solution now has three possible forms. If $b^2 > 4ac$, then Equation B.4 yields two real roots s_1 and s_2 and the complementary function is

$$x = Ae^{s_1 t} + Be^{s_2 t} \tag{B.5}$$

where A and B are arbitrary constants. If $b^2 = 4ac$, then Equation B.4 gives only a single value of s and the complementary function is

$$x = (A + Bt)e^{st} \tag{B.6}$$

If $b^2 < 4ac$, then Equation B.4 yields two complex roots:

$$s_1 = \frac{-b}{2a} + \frac{\sqrt{b^2-4ac}}{2a}\,i = p+qi, \text{ say, and } s_2 = p-qi \tag{B.7}$$

and the complementary function is

$$x = Ae^{(p+qi)t} + Be^{(p-qi)t} \tag{B.8}$$

Using Euler's equation,

$$e^{it} = \cos t + i\sin t \tag{B.9}$$

we can write Equation B.8 as

$$x = e^{pt}(C\sin qt + D\cos qt) \tag{B.10}$$

where C and D are arbitrary constants.

Of course, this third case is the one we are normally dealing with in dynamics because it is the only one of the three that results in a solution that oscillates with time.

The particular integral is normally found by substituting a trial solution having a similar form to the function $F(t)$. For example, if $F(t) = Rt + S$ where R and S are constants, we try a particular integral of the form

$$x(t) = d_1 t + d_2 \rightarrow \dot{x} = d_1,\; \ddot{x} = 0$$

Substituting into Equation B.1 gives

$$a(0) + b(d_1) + c(d_1 t + d_2) = Rt + S$$

Equating coefficients of powers of t on either side of the equation gives $d_1 = R/c$ and $d_2 = S/c - bR/c^2$.

Example B.1

Solve $\dfrac{d^2x}{dt^2} + 2\dfrac{dx}{dt} + 5x = \sin t$, subject to the initial conditions $x = \dfrac{dx}{dt} = 0$ at $t = 0$.

The auxiliary equation is

$$s^2 + 2s + 5 = 0$$

which has roots $s_1 = -1 + 2i$, $s_2 = -1 - 2i$, giving the complementary function

$$x = e^{-t}(C\sin 2t + D\cos 2t)$$

Given the sinusoidal forcing function, try a particular integral of the form

$$x = d_1\sin t + d_2\cos t$$

Differentiating and substituting into the ODE:

$$(-d_1 \sin t - d_2 \cos t) + 2(d_1 \cos t - d_2 \sin t) + 5(d_1 \sin t + d_2 \cos t) = \sin t$$

Equating the sine and cosine coefficients on either side of the equation gives $d_1 = 0.2$ and $d_2 = -0.1$. The general solution (complementary function plus particular integral) is then

$$x = e^{-t}(C \sin 2t + D \cos 2t) + 0.2 \sin t - 0.1 \cos t$$

Finally, the constants C and D can be found from the initial conditions. Substituting $t = 0$, $x = 0$ into the general solution gives $D = 0.1$, while $t = 0$, $dx/dt = 0$ gives $C = -0.05$. The full solution can now be written as

$$x = e^{-t}(-0.05 \sin 2t + 0.1 \cos 2t) + 0.2 \sin t - 0.1 \cos t$$

We saw in Appendix A that, as an alternative to writing sine and cosine terms at a particular frequency, we can express the result just as a sine term, but offset by a phase angle. Thus, by comparison with Equations A.5 through A.7, our solution could also be written as

$$x = 0.112e^{-t} \sin(2t - 1.107) + 0.224 \sin(t - 0.464)$$

Example B.2

Solve the equation for free vibration of a damped single-degree-of-freedom (SDOF) system:

$$\ddot{x} + 2\xi\omega_n\dot{x} + \omega_n^2 x = 0$$

taking the initial displacement as X and the initial velocity as zero. Consider all three possible solutions.

The auxiliary equation is

$$s^2 + 2\xi\omega_n s + \omega_n^2 = 0$$

which has the solutions:

$$s = \omega_n\left(-\xi \pm \sqrt{\xi^2 - 1}\right)$$

Three different solutions are possible, depending on the value of ξ, which can result in the argument of the square root term being positive, zero or negative. Consider each case in turn.

(a) $\xi > 1$. There are two real roots for s, so we can write the solution as

$$x = Ae^{\left(-\xi + \sqrt{\xi^2 - 1}\right)\omega_n t} + Be^{\left(-\xi - \sqrt{\xi^2 - 1}\right)\omega_n t}$$

Differentiating:

$$\dot{x} = A\left(-\xi + \sqrt{\xi^2 - 1}\right)\omega_n e^{\left(-\xi + \sqrt{\xi^2 - 1}\right)\omega_n t} + B\left(-\xi - \sqrt{\xi^2 - 1}\right)\omega_n e^{\left(-\xi - \sqrt{\xi^2 - 1}\right)\omega_n t}$$

Then, inserting the initial conditions:

$$x(0) = X = A + B$$

$$\dot{x}(0) = 0 = A\left(-\xi + \sqrt{\xi^2 - 1}\right)\omega_n + B\left(-\xi - \sqrt{\xi^2 - 1}\right)\omega_n$$

From which we find:

$$A = X\frac{\left(\xi + \sqrt{\xi^2 - 1}\right)}{2\sqrt{\xi^2 - 1}}, \ B = X\frac{\left(-\xi + \sqrt{\xi^2 - 1}\right)}{2\sqrt{\xi^2 - 1}}$$

and the full solution is therefore:

$$x = \frac{X}{2\sqrt{\xi^2 - 1}}\left[\left(\xi + \sqrt{\xi^2 - 1}\right)e^{\left(-\xi + \sqrt{\xi^2 - 1}\right)\omega_n t} + \left(-\xi + \sqrt{\xi^2 - 1}\right)e^{\left(-\xi - \sqrt{\xi^2 - 1}\right)\omega_n t}\right]$$

(b) $\xi = 1$. There is one real root for s, so we can write the solution as

$$x = (A + Bt)e^{-\omega_n t}$$

Differentiating:

$$\dot{x} = -\omega_n(A + Bt)e^{-\omega_n t} + Be^{-\omega_n t}$$

Applying the initial conditions:

$$x(0) = X = A$$

$$\dot{x}(0) = 0 = -\omega_n A + B \ \rightarrow \ B = \omega_n X$$

So the full solution can be written as

$$x = X(1 + \omega_n t)e^{-\omega_n t}$$

(c) $\xi < 1$. There are two complex roots for s, giving a solution of the form:

$$x = e^{-\xi \omega_n t}\left(A\sin\omega_n\sqrt{1 - \xi^2}t + B\cos\omega_n\sqrt{1 - \xi^2}t\right)$$

We can solve this in the form given or we can recast it in terms of a phase angle as in Appendix A, Equation A.6. Here we will take the latter approach. If we also simplify the algebra by making the substitution

$$\omega_d = \omega_n \sqrt{1 - \xi^2}$$

then the solution becomes

$$x = Ce^{-\xi \omega_n t} \sin(\omega_d t + \phi)$$

Differentiating gives

$$\dot{x} = -C\xi\omega_n e^{-\xi \omega_n t} \sin(\omega_d t + \phi) + C\omega_d e^{-\xi \omega_n t} \cos(\omega_d t + \phi)$$

Applying the initial conditions then leads to

$$x(0) = X = C \sin\phi$$

$$\dot{x}(0) = 0 = -C\xi\omega_n \sin\phi + C\omega_d \cos\phi$$

Solving these equations simultaneously gives

$$C = \frac{X}{\sqrt{1 - \xi^2}}, \quad \tan\phi = \frac{\sqrt{1 - \xi^2}}{\xi}$$

and hence the final solution is

$$x = \frac{X}{\sqrt{1 - \xi^2}} e^{-\xi \omega_n t} \sin(\omega_d t + \phi)$$

Appendix C: Fourier analysis

The ability to express time-domain functions or signals in terms of their frequency components stems from the work of Joseph Fourier (1768–1830), whose principal interest was heat flow in solids. The initial idea of representing periodic functions by sets of sine and cosine terms gave rise to the Fourier series, and this has since been extended to the Fourier transform, which can be applied to non-periodic signals as well.

C.1 FOURIER SERIES

Fourier showed that any periodic function could be represented as the sum of a series of sine and/or cosine terms. The first term in the series has the same frequency as the function being simulated, the subsequent terms being higher harmonics, that is, integer multiples of the base frequency. The general form of the Fourier series for some periodic function $x(t)$ with circular frequency ω_0 and period $T = 2\pi/\omega_0$ is

$$x(t) = \frac{A_0}{2} + A_1 \cos(\omega_0 t) + A_2 \cos(2\omega_0 t) \ldots + B_1 \sin(\omega_0 t) + B_2 \sin(2\omega_0 t) \ldots$$

$$= \frac{A_0}{2} + \sum_{n=1}^{\infty} [A_n \cos(n\omega_0 t) + B_n \sin(n\omega_0 t)] \tag{C.1}$$

where the coefficients are given by

$$A_n = \frac{2}{T} \int_{-T/2}^{T/2} x(t) \cos(n\omega_0 t) \, dt, \quad B_n = \frac{2}{T} \int_{-T/2}^{T/2} x(t) \sin(n\omega_0 t) \, dt \tag{C.2}$$

Simpler forms can be obtained for *even* and *odd* functions. An even function is one that is symmetric about $t = 0$, that is, $f(-t) = f(t)$, while an odd function displays anti-symmetry about $t = 0$, that is, $f(-t) = -f(t)$. It is easy to show that

- An *odd* function can be represented by a Fourier *sine* series (since sine is itself an odd function), that is, $A_n = 0$ for all n, and
- An *even* function can be represented by a Fourier *cosine* series, that is, $B_n = 0$ for all n.

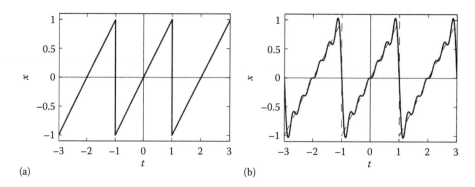

(a) (b)

Figure C.1 (a) Saw tooth wave and (b) its Fourier series approximation (first six terms).

Example C.1

Find the Fourier series for the saw tooth wave shown in Figure C.1a.

This is an odd function and so can be represented by a sine series. The period $T = 2$ s and circular frequency $\omega_0 = 2\pi/T = \pi$ rad/s. Noting that $\sin(n\pi) = \sin(-n\pi) = 0$ and $\cos(n\pi) = \cos(-n\pi) = (-1)^n$, the Fourier coefficients are

$$B_n = \frac{2}{T}\int_{-T/2}^{T/2} x(t)\sin(n\omega_0 t)\,dt = \frac{2}{2}\int_{-1}^{1} t\sin(n\pi t)\,dt = \left[\frac{\sin(n\pi t)}{n^2\pi^2} - \frac{t\cos(n\pi t)}{n\pi}\right]_{-1}^{1}$$

$$= \left[\frac{\sin(n\pi) - \sin(-n\pi)}{n^2\pi^2} - \frac{\cos(n\pi) + \cos(-n\pi)}{n\pi}\right] = \frac{-2(-1)^n}{n\pi}$$

Figure C.1b shows the level of agreement achieved by Fourier terms up to $n = 6$.

Fourier series for some simple waveforms are shown in Table C.1.

C.2 COMPLEX FOURIER SERIES AND FOURIER TRANSFORM

The Fourier series can also be expressed in terms of the complex exponential. Recall Euler's formula, Equation A.8:

$$e^{i\omega_0 t} = \cos\omega_0 t + i\sin\omega_0 t$$

Noting that cosine is an even function and sine odd, we can also write

$$e^{-i\omega_0 t} = \cos(-\omega_0 t) + i\sin(-\omega_0 t) = \cos\omega_0 t - i\sin\omega_0 t$$

Rearranging these two expressions gives

$$\cos\omega_0 t = \frac{1}{2}(e^{i\omega_0 t} + e^{-i\omega_0 t}), \quad \sin\omega_0 t = -\frac{i}{2}(e^{i\omega_0 t} - e^{-i\omega_0 t})$$

Table C.1 Fourier series of some periodic functions

Waveform	Series
	$x(t) = \dfrac{4A}{\pi}\left(\cos \omega_0 t - \dfrac{1}{3}\cos 3\omega_0 t + \dfrac{1}{5}\cos 5\omega_0 t \ldots\right)$
	$x(t) = \dfrac{8A}{\pi^2}\left(\cos \omega_0 t + \dfrac{1}{9}\cos 3\omega_0 t + \dfrac{1}{25}\cos 5\omega_0 t \ldots\right)$
	$x(t) = \dfrac{2A}{\pi}\left(\sin \omega_0 t - \dfrac{1}{2}\sin 2\omega_0 t + \dfrac{1}{3}\sin 3\omega_0 t \ldots\right)$
	$x(t) = \dfrac{2A}{\pi}\left(\sin \omega_0 t + \dfrac{1}{2}\sin 2\omega_0 t + \dfrac{1}{3}\sin 3\omega_0 t \ldots\right)$
	$x(t) = \dfrac{A}{\pi}\left(1 + \dfrac{\pi}{2}\cos \omega_0 t + \dfrac{2}{3}\cos 2\omega_0 t - \dfrac{2}{15}\cos 4\omega_0 t \ldots\right)$

Note: All functions have amplitude A, period T and $\omega_0 = 2\pi/T$.

and substituting into Equation C.1 gives

$$x(t) = \frac{A_0}{2} + \sum_{n=1}^{\infty}\left(\frac{1}{2}(A_n - iB_n)e^{in\omega_0 t} + \frac{1}{2}(A_n + iB_n)e^{-in\omega_0 t}\right)$$

$$= \frac{A_0}{2} + \sum_{n=1}^{\infty}\left(C_n e^{in\omega_0 t} + C_n^* e^{-in\omega_0 t}\right) \qquad (C.3)$$

$$= \sum_{n=-\infty}^{\infty} C_n e^{in\omega_0 t}$$

where

$$C_n = \frac{1}{2}(A_n - iB_n) = \frac{1}{T}\int_{-T/2}^{T/2} x(t)e^{-in\omega_0 t}\, dt \qquad (C.4)$$

Equations C.3 and C.4 define the complex Fourier series, which may be used as an alternative to the sine and cosine form introduced earlier.

Table C.2 Fourier transforms of some common functions

Description	$x(t)$	$X(\omega)$
Unit impulse at time $t = 0$	$\delta(t)$	1
Unit step at time $t = 0$	$u(t) = \begin{cases} 0 & t<0 \\ 1 & t>0 \end{cases}$	$\pi\delta(\omega) + \dfrac{1}{i\omega}$
One-sided exponential	$u(t)e^{-\alpha t}\ (\alpha > 0)$	$\dfrac{1}{\alpha + i\omega}$
Two-sided exponential	$e^{-\alpha|t|}\ (\alpha > 0)$	$\dfrac{2\alpha}{\alpha^2 + \omega^2}$
Complex exponential	$e^{-i\omega_0 t}$	$2\pi\delta(\omega - \omega_0)$
Sinusoids	$\cos \omega_0 t$	$\pi[\delta(\omega - \omega_0) + \delta(\omega + \omega_0)]$
	$\sin \omega_0 t$	$i\pi[-\delta(\omega - \omega_0) + \delta(\omega + \omega_0)]$
Rectangular pulse		$a\,\dfrac{\sin(\omega a/2)}{\omega a/2}$
Triangular pulse		$a\,\dfrac{\sin^2(\omega a/2)}{(\omega a/2)^2}$

We now consider how the complex Fourier series can be used to derive the Fourier transform, enabling the representation in the frequency domain of functions which need not be periodic. First, substituting for C_n allows us to write the single expression:

$$x(t) = \sum_{n=-\infty}^{\infty} \frac{1}{T} \int_{-T/2}^{T/2} x(t)e^{-in\omega_0 t}\,\mathrm{d}t\, e^{in\omega_0 t} \tag{C.5}$$

Next we note that the Fourier series produces terms with frequency $\omega_0, 2\omega_0, 3\omega_0 \ldots$ where $\omega_0 = 2\pi/T$. The increment between frequencies is therefore $\Delta\omega = 2\pi/T$. Using this to eliminate T from Equation C.5 gives

$$x(t) = \sum_{n=-\infty}^{\infty} \frac{\Delta\omega}{2\pi} \int_{-T/2}^{T/2} x(t)e^{-in\omega_0 t}\,\mathrm{d}t\, e^{in\omega_0 t} \tag{C.6}$$

If we now let $T \to \infty$, then the increment $\Delta\omega$ becomes the infinitesimal quantity $\mathrm{d}\omega$, the points $n\omega_0$ become the continuous variable ω and the summation becomes an integral between infinite limits:

$$x(t) = \int_{-\infty}^{\infty} \frac{\mathrm{d}\omega}{2\pi} \int_{-\infty}^{\infty} x(t)e^{-i\omega t}\,\mathrm{d}t\, e^{i\omega t} \tag{C.7}$$

This can be written as

$$x(t) = \frac{1}{2\pi} \int_{-\infty}^{\infty} X(\omega)e^{i\omega t}\, d\omega \tag{C.8}$$

where

$$X(\omega) = \int_{-\infty}^{\infty} x(t)e^{-i\omega t}\, dt \tag{C.9}$$

Equation C.9 is the Fourier transform and Equation C.8 is the inverse Fourier transform, or Fourier integral. Together, they are known as the Fourier transform pair, allowing a signal to be switched between a function of time and one of frequency. Note that, in splitting Equation C.7 into the two Equations C.8 and C.9, the factor of $1/2\pi$ could have been placed in either equation; the choice is arbitrary and there is no consensus on this point. Here it is placed in the inverse Fourier transform equation, which is believed to be the most common choice.

The Fourier transforms of some common functions are listed in Table C.2.

Appendix D: Laplace transforms

The solution of differential equations can often be simplified by converting them to algebraic equations in terms of the complex Laplace variable s. Expressing the governing equations of dynamic systems in terms of s rather than time also makes them amenable to representation in block diagram form, which offers advantages in many cases.

The Laplace transform of a function $x(t)$ is given by

$$\bar{x}(s) = \int_0^\infty e^{-st} x(t)\,dt \tag{D.1}$$

If $x(t)$ has a simple functional form, this can be evaluated quite easily, as in the following examples.

Example D.1

Find the Laplace transform of the unit step function:

$$x(t) = \begin{cases} 0 & t < 0 \\ 1 & t > 0 \end{cases}$$

$x(t)$ takes the value of 1 over the entire range of the Laplace transform integral (Equation D.1), so:

$$\bar{x}(s) = \int_0^\infty e^{-st}.1\,dt = \left[\frac{-1}{s} e^{-st} \right]_0^\infty = \frac{1}{s}$$

Example D.2

Repeat the calculation in Example D.1, but with the step occurring at time $t = a$, that is,

$$x(t) = \begin{cases} 0 & t < a \\ 1 & t > a \end{cases}$$

$x(t)$ is now discontinuous over the range of t from zero to infinity, so we must split the integral in two:

$$\bar{x}(s) = \int_0^a e^{-st}.0\,dt + \int_a^\infty e^{-st}.1\,dt = \left[\frac{-1}{s}e^{-st}\right]_a^\infty = \frac{1}{s}e^{-as}$$

Therefore, a time shift in the time domain causes a scaling by a negative exponential in the Laplace domain.

Example D.3

Find the Laplace transform of $x(t) = e^{-\alpha t}$.

$$\bar{x}(s) = \int_0^\infty e^{-st}e^{-\alpha t}\,dt = \int_0^\infty e^{-(s+\alpha)t}\,dt = \left[\frac{-1}{s+\alpha}e^{-(s+\alpha)t}\right]_0^\infty = \frac{1}{s+\alpha}$$

As the function gets more complicated, so does the evaluation of its Laplace transform. Rather than slog through some tedious integration, it is easier and quicker to use a look-up table such as Table D.1.

Some important properties of Laplace transforms are

Scaling: If $x(t) \Rightarrow \bar{x}(s)$, then $b.x(t) \Rightarrow b.\bar{x}(s)$ where b is a constant.

Addition: If $x(t) \Rightarrow \bar{x}(s)$ and $y(t) \Rightarrow \bar{y}(s)$, then $x(t) + y(t) \Rightarrow \bar{x}(s) + \bar{y}(s)$.

Shifting: It is easy to show that the result in Example D.2 is a general one, that is, if $x(t) \Rightarrow \bar{x}(s)$ then:

$$x(t-a) \Rightarrow \bar{x}(s).e^{-as} \tag{D.2}$$

and conversely

$$x(t).e^{at} \Rightarrow \bar{x}(s-a) \tag{D.3}$$

Differentiation: If $x(t) \Rightarrow \bar{x}(s)$ and $x(t)$ has initial conditions $x(0)$ and $\dot{x}(0)$, then:

$$\dot{x}(t) \Rightarrow s\bar{x}(s) - x(0) \tag{D.4}$$

and

$$\ddot{x}(t) \Rightarrow s^2\bar{x}(s) - sx(0) - \dot{x}(0) \tag{D.5}$$

Table D.I Laplace transforms of common functions

Description	$x(t)$	$\bar{x}(s)$
Unit impulse at time $t = 0$	$\delta(t)$	1
Unit step at time $t = 0$	$\begin{cases} 0 & t < 0 \\ 1 & t > 0 \end{cases}$	$\dfrac{1}{s}$
Powers of t	t	$\dfrac{1}{s^2}$
	t^2	$\dfrac{2}{s^3}$
	t^n	$\dfrac{n!}{s^{n+1}}$
Exponential functions:	$e^{-\alpha t}$	$\dfrac{1}{s+\alpha}$
	$te^{-\alpha t}$	$\dfrac{1}{(s+\alpha)^2}$
	$1 - e^{-\alpha t}$	$\dfrac{\alpha}{s(s+\alpha)}$
Sinusoids	$\sin \omega t$	$\dfrac{\omega}{s^2+\omega^2}$
	$\cos \omega t$	$\dfrac{s}{s^2+\omega^2}$
	$t \sin \omega t$	$\dfrac{2\omega s}{(s^2+\omega^2)^2}$
	$t \cos \omega t$	$\dfrac{s^2-\omega^2}{(s^2+\omega^2)^2}$
	$1 - \cos \omega t$	$\dfrac{\omega^2}{s(s^2+\omega^2)}$
Damped impulse response	$\dfrac{1}{\omega_d} e^{-\xi \omega_n t} \sin \omega_d t$	$\dfrac{1}{s^2+2\xi\omega_n s+\omega_n^2}$
	where $\omega_d = \omega_n \sqrt{1-\xi^2}$	

Example D.4

Solve:

$$\frac{d^2x}{dt^2} + 2\frac{dx}{dt} + 5x = \sin t,$$

subject to the initial conditions $x = \dfrac{dx}{dt} = 0$ at $t = 0$.

(This equation was solved by the conventional method in Example B.1.)

Start by taking the Laplace transform of the equation. Note that, when taking the transform of the differentials, the initial conditions (zero in this case) are immediately included in the solution.

$$s^2\bar{x}(s) + 2s\bar{x}(s) + 5\bar{x}(s) = \frac{1}{s^2 + 1}$$

This is easily rearranged to give

$$\bar{x}(s) = \frac{1}{(s^2 + 1)(s^2 + 2s + 5)}$$

Before we can invert this to find $x(t)$ we need to break it down into separate terms that look like ones in Table D.1. First, split into two partial fractions:

$$\frac{1}{(s^2 + 1)(s^2 + 2s + 5)} = \frac{As + B}{(s^2 + 1)} + \frac{Cs + D}{(s^2 + 2s + 5)}$$

$$= \frac{(As + B)(s^2 + 2s + 5) + (Cs + D)(s^2 + 1)}{(s^2 + 1)(s^2 + 2s + 5)}$$

Equating powers of s in the left- and right-hand numerators then gives $A = -0.1$, $B = 0.2$, $C = 0.1$, $D = 0$. So:

$$\bar{x}(s) = -0.1\frac{s}{(s^2 + 1)} + 0.2\frac{1}{(s^2 + 1)} + 0.1\frac{s}{(s^2 + 2s + 5)}$$

The first two terms are now in an easily invertible form but the third needs some further manipulation. Write it as a function of $(s + 1)$ in anticipation of using the shifting rule (Equation D.3):

$$\frac{s}{(s^2 + 2s + 5)} = \frac{s}{(s+1)^2 + 2^2} = \frac{s+1}{(s+1)^2 + 2^2} - 0.5\frac{2}{(s+1)^2 + 2^2}$$

The full expression can now be written as

$$\bar{x}(s) = -0.1\frac{s}{(s^2 + 1)} + 0.2\frac{1}{(s^2 + 1)} + 0.1\frac{s+1}{(s+1)^2 + 2^2} - 0.05\frac{2}{(s+1)^2 + 2^2}$$

Finally, this is in a form that can be inverted. Comparing with Table D.1, the first two terms invert to simple cosine and sine functions, while the second two also give cosine and sine functions but scaled by an exponential, using the shifting rule (Equation D.3).

$$x(t) = -0.1\cos t + 0.2\sin t + 0.1e^{-t}\cos 2t - 0.05e^{-t}\sin 2t$$

It can be seen that this is the same expression as in Example B.1.

Appendix E: Eigenvalues and eigenvectors of matrices

For a square (n rows $\times$ n columns) matrix $\mathbf{A}$, the eigenvalues λ and eigenvectors $\mathbf{u}$ are the solutions of the equation

$$\mathbf{Au} = \lambda\mathbf{u} \qquad\qquad\qquad\qquad\qquad\text{(E.1)}$$

There are n solutions to Equation E.1, each yielding an eigenvalue-eigenvector pair. The eigenvectors have dimensions ($n \times 1$) and the scaling of their elements is arbitrary – what matters is the ratios between the elements.

The most obvious way to solve Equation E.1 is to rewrite it as

$$(\mathbf{A} - \lambda\mathbf{I})\,\mathbf{u} = 0 \qquad\qquad\qquad\qquad\qquad\text{(E.2)}$$

where $\mathbf{I}$ is a unit matrix of dimension n. It is then clear that, unless $\mathbf{u}$ is always zero, the matrix defined by the bracketed term must have no inverse. In other words, its determinant must be zero. This enables determination of the eigenvalues λ, and the eigenvectors $\mathbf{u}$ can then be found by back-substitution into Equation E.2.

The determinant of a 2 $\times$ 2 matrix can be found from:

$$\mathbf{A} = \begin{bmatrix} a_1 & a_2 \\ b_1 & b_2 \end{bmatrix} \rightarrow \det \mathbf{A} = a_1 b_2 - b_1 a_2 \qquad\qquad\qquad\text{(E.3)}$$

and that of a 3 $\times$ 3 matrix from:

$$\mathbf{A} = \begin{bmatrix} a_1 & a_2 & a_3 \\ b_1 & b_2 & b_3 \\ c_1 & c_2 & c_3 \end{bmatrix} \rightarrow \det \mathbf{A} = a_1(b_2 c_3 - c_2 b_3) - a_2(b_1 c_3 - c_1 b_3) + a_3(b_1 c_2 - c_1 b_2) \qquad\text{(E.4)}$$

For matrices larger than 3 $\times$ 3, evaluation of the determinant becomes cumbersome; many software packages incorporate far more efficient ways of finding the eigenvalues.

Example E.1

Find the eigenvalues and eigenvectors of $\mathbf{A} = \begin{bmatrix} 5 & -2 \\ -2 & 2 \end{bmatrix}$

Write $\mathbf{A} - \lambda\mathbf{I} = \begin{bmatrix} 5-\lambda & -2 \\ -2 & 2-\lambda \end{bmatrix}$, then put $\det(\mathbf{A} - \lambda\mathbf{I}) = 0$:

$$(5-\lambda)(2-\lambda) - (-2)(-2) = 0$$
$$\lambda^2 - 7\lambda + 6 = 0$$
$$(\lambda - 1)(\lambda - 6) = 0$$

Therefore, the two solutions are $\lambda_1 = 1$, $\lambda_2 = 6$. Substituting each in turn into Equation E.2:

$$\lambda_1 : (\mathbf{A} - \lambda_1\mathbf{I})\mathbf{u}_1 = \begin{bmatrix} 4 & -2 \\ -2 & 1 \end{bmatrix} \begin{bmatrix} u_1 \\ u_2 \end{bmatrix} = 0 \rightarrow \mathbf{u}_1 = \frac{1}{\sqrt{5}} \begin{bmatrix} 1 \\ 2 \end{bmatrix}$$

$$\lambda_2 : (\mathbf{A} - \lambda_2\mathbf{I})\mathbf{u}_2 = \begin{bmatrix} -1 & -2 \\ -2 & -4 \end{bmatrix} \begin{bmatrix} u_1 \\ u_2 \end{bmatrix} = 0 \rightarrow \mathbf{u}_2 = \frac{1}{\sqrt{5}} \begin{bmatrix} -2 \\ 1 \end{bmatrix}$$

Note that in both cases the two lines of the matrix equation give the same relationship between the two elements of $\mathbf{u}$; this is why it is only possible to find a ratio between them, not absolute values. Here, we have scaled the values so that $\mathbf{u}^T\mathbf{u} = 1$; while any scaling is acceptable, it will soon become clear why this is a useful thing to do.

Eigenvectors have some very useful properties that can be used to simplify matrix equations. The eigenvectors of a symmetric matrix (i.e. one for which $\mathbf{A}^T = \mathbf{A}$) are orthogonal to each other, meaning that, if we multiply two different eigenvectors together, the result is zero. Using the scaling introduced in Example E.1, we can therefore write

$$\mathbf{u}_i^T\mathbf{u}_j = 0 \text{ if } i \neq j$$
$$\mathbf{u}_i^T\mathbf{u}_j = 1 \text{ if } i = j$$

(E.5)

It is easy to see that this is true for the two eigenvectors in Example E.1.

If we now create a matrix $\mathbf{U}$, comprising all the eigenvectors of $\mathbf{A}$ written as column vectors, then $\mathbf{U}^T\mathbf{U}$ is a unit matrix, that is, all its elements are zero except those on the leading diagonal, which are 1. Suppose we now reformulate Equation E.1 using the matrix of all eigenvectors $\mathbf{U}$ rather than a single eigenvector $\mathbf{u}$:

$$\mathbf{A}\mathbf{U} = \mathbf{\Lambda}\mathbf{U} \quad \text{where} \quad \mathbf{\Lambda} = \begin{bmatrix} \lambda_1 & & & \\ & \lambda_2 & & \\ & & \ddots & \\ & & & \lambda_n \end{bmatrix}$$

(E.6)

Pre-multiply by $\mathbf{U}^T$, noting that $\mathbf{U}^T\mathbf{U} = \mathbf{I}$:

$$\mathbf{U}^T\mathbf{A}\mathbf{U} = \mathbf{U}^T\mathbf{\Lambda}\mathbf{U} = \mathbf{\Lambda}(\mathbf{U}^T\mathbf{U}) = \mathbf{\Lambda}$$

(E.7)

Putting Equation E.7 into words, pre- and post-multiplying $\mathbf{A}$ by its matrix of eigenvectors (appropriately normalised) converts $\mathbf{A}$ to a matrix whose leading diagonal elements are the eigenvalues, and whose other elements are all zero.

Example E.2

Diagonalise the matrix $\mathbf{A} = \begin{bmatrix} 5 & -2 \\ -2 & 2 \end{bmatrix}$

The eigenvectors of this matrix are given in Example E.1. Writing them in matrix form:

$$\mathbf{U} = \frac{1}{\sqrt{5}} \begin{bmatrix} 1 & -2 \\ 2 & 1 \end{bmatrix}$$

Then Equation E.7 gives

$$\mathbf{U}^{\mathsf{T}}\mathbf{A}\mathbf{U} = \frac{1}{\sqrt{5}} \begin{bmatrix} 1 & 2 \\ -2 & 1 \end{bmatrix} \begin{bmatrix} 5 & -2 \\ -2 & 2 \end{bmatrix} \frac{1}{\sqrt{5}} \begin{bmatrix} 1 & -2 \\ 2 & 1 \end{bmatrix}$$

$$= \frac{1}{5} \begin{bmatrix} 1 & 2 \\ -2 & 1 \end{bmatrix} \begin{bmatrix} 1 & -12 \\ 2 & 6 \end{bmatrix}$$

$$= \frac{1}{5} \begin{bmatrix} 5 & 0 \\ 0 & 30 \end{bmatrix} = \begin{bmatrix} 1 & 0 \\ 0 & 6 \end{bmatrix}$$

As expected, the diagonal values are the eigenvalues of **A**. We went through the matrix multiplication here to show that it works, but in general it would be acceptable simply to jump straight to the answer, knowing the eigenvalues.

Appendix F: Partial differential equations

F.1 DIFFERENTIATION OF FUNCTIONS OF MORE THAN ONE VARIABLE

Many engineering quantities vary as functions of more than one independent variable, such as position in two dimensions, or position and time. In order to find their rate of change as one of the independent variables changes, we can differentiate with respect to that variable while treating terms involving the other variables as though they were constants. To distinguish a partial differential from an ordinary one, we denote it by a curly ∂. For example, for a simple function of two variables:

$$x(y,z) = (1-y^2)\sin\frac{\pi z}{2} \tag{F.1}$$

the two partial differentials are

$$\frac{\partial x}{\partial y} = -2y\sin\frac{\pi z}{2}, \quad \frac{\partial x}{\partial z} = (1-y^2)\frac{\pi}{2}\cos\frac{\pi z}{2} \tag{F.2}$$

A somewhat more complex function of three variables

$$g(y,z,t) = y\cos yz \sin \omega t \tag{F.3}$$

results in three partial differentials:

$$\frac{\partial g}{\partial y} = (\cos yz - yz\sin yz)\sin\omega t, \quad \frac{\partial g}{\partial z} = -y^2\sin yz\sin\omega t, \quad \frac{\partial g}{\partial t} = y\omega\cos yz\cos\omega t \tag{F.4}$$

Higher-order differentials can be obtained by repeated differentiation. For example, the expressions in Equation F.2 can be differentiated again to give

$$\frac{\partial^2 x}{\partial y^2} = -2\sin\frac{\pi z}{2}, \quad \frac{\partial^2 x}{\partial z^2} = -(1-y^2)\frac{\pi^2}{4}\sin\frac{\pi z}{2} \tag{F.5}$$

However, another second differential, known as the cross-differential, is also possible. If we differentiate the first expression in Equation F.2 with respect to z, or the second term with respect to y, we get:

$$\frac{\partial^2 x}{\partial y \partial z} = -\pi y \cos \frac{\pi z}{2} \qquad\qquad (F.6)$$

The physical meanings of most of the partial differentials are quite straightforward. For instance, $\partial x/\partial y$ is the slope of function x as we move in the y-direction, at a fixed z and $\partial^2 x/\partial y^2$ is the curvature (rate of change of slope) of x with y, at fixed z. The cross-differential is a little harder to visualise; $\partial^2 x/\partial y \partial z$ is the rate of change of the y-direction slope as we move in the z-direction, at fixed y (or *vice versa*).

F.2 SOLUTION BY SEPARATION OF VARIABLES

Partial differentials can be combined into equations in much the same way as ordinary differentials. For instance, in beam vibrations the displacement x is a function of both position along the beam, z, and time, t. For a uniform beam, the governing equation for free vibrations relates the fourth differential with respect to position to the second differential with respect to time according to

$$EI \frac{\partial^4 x}{\partial z^4} + m \frac{\partial^2 x}{\partial t^2} = 0 \qquad\qquad (F.7)$$

A well-known solution procedure for equations of this sort is the method of separation of variables. We assume that the function $x(z,t)$ can be written as a product of two single-variable functions, one in terms of z and one in terms of t:

$$x(z,t) = u(z).y(t) \qquad\qquad (F.8)$$

When differentiating Equation F.8 with respect to z, the time function is simply treated as a constant coefficient, and when differentiating with respect to t, the position function is treated likewise. Using a dash to denote differentiation with respect to z and a dot for differentiation with respect to t, the differentials we need are therefore:

$$x'''' = u''''.y, \quad \ddot{x} = u.\ddot{y} \qquad\qquad (F.9)$$

Substituting into Equation F.7 and dividing through by $EIuy$:

$$\frac{u''''}{u} = -\frac{m}{EI} \frac{\ddot{y}}{y} \qquad\qquad (F.10)$$

The left-hand side is a function solely of z and the right-hand side solely of t. The only way the two sides can be equal for all z and t is if they both equal the same, constant value. For reasons that will become clear in a moment, we want the constant to be positive, and it will turn out to be convenient to write it as α^4. Equation F.10 can then be written as

$$\frac{u''''}{u} = -\frac{m}{EI}\frac{\ddot{y}}{y} = \alpha^4$$

which can be separated into two ordinary differential equations:

$$u'''' - \alpha^4 u = 0 \tag{F.11}$$

and

$$\ddot{y} + \frac{\alpha^4 m}{EI} y = 0 \tag{F.12}$$

The reason why the constant must be positive should now be clear. This is a vibration problem, for which we want the displacement to vary with time on an oscillatory way; if the coefficient of y in Equation F.12 were negative, an exponential solution would result, and there would be no vibration. The solution to Equation F.12 was derived in Appendix B. It is:

$$y(t) = A_1 \sin \omega_n t + A_2 \cos \omega_n t \tag{F.13}$$

where

$$\omega_n^2 = \frac{\alpha^4 EI}{m} \quad \text{or} \quad \alpha = \left(\frac{\omega_n^2 m}{EI}\right)^{1/4} \tag{F.14}$$

To solve Equation F.11 we adopt the usual form of trial solution:

$$u(z) = Ce^{sz} \tag{F.15}$$

Substituting into Equation F.11 leads to

$$(s^4 - \alpha^4)\, Ce^{sz} = 0 \tag{F.16}$$

which has four roots: $s = \alpha, -\alpha, i\alpha, -i\alpha$. Equation F.15 then gives

$$u(z) = C_1 e^{\alpha z} + C_2 e^{-\alpha z} + C_3 e^{i\alpha z} + C_4 e^{-i\alpha z} \tag{F.17}$$

But noting that:

$$e^z = \cosh z + \sinh z, \quad e^{iz} = \cos z + i \sin z \tag{F.18}$$

this is usually written in the alternative form:

$$u(z) = B_1 \sin \alpha z + B_2 \cos \alpha z + B_3 \sinh \alpha z + B_4 \cosh \alpha z \qquad \text{(F.19)}$$

Finally, the full solution is obtained by taking the product of the solutions for y and u:

$$x(z,t) = (A_1 \sin \omega_n t + A_2 \cos \omega_n t)(B_1 \sin \alpha z + B_2 \cos \alpha z + B_3 \sinh \alpha z + B_4 \cosh \alpha z) \quad \text{(F.20)}$$

The final solution requires the application of the appropriate boundary and initial conditions. Imposition of four boundary conditions will enable relationships to be determined between the coefficients B_1 to B_4, and will yield multiple solutions for the parameter α, say α_i, where $i = 1, 2, 3...$ These solutions, in turn, will lead to multiple natural frequencies ω_i from Equation F.14. Finally, the coefficients A_1 and A_2 can be found from the initial conditions.

Appendix G: Tutorial problems
Answers and hints

Hints and comments are in square brackets.

CHAPTER I

1. Amplitude = 13 mm, phase = 22.6° (or 0.39 rads)

2. Floor: 0.32 mm, roof: 1.30 mm.

3. $z = 5e^{0.93i}$ [Essential to work in radians, not degrees, when using this form.]

4. $24.6\sin(\omega t - 0.12)$ (where 0.12 rads = 6.9°) [Easiest method is to use phasors.]

5. $c = 318.3$ Ns/m. At 5 Hz, damper force = 100 N [Since force is proportional to velocity, which in turn is proportional to frequency.]

6. $k = 18.3 \times 10^6$ N/m. [The stiffness of the columns can be deduced from the deformation shape – splitting a column at mid-height, each half deforms as a tip-loaded cantilever and the total lateral deflection is the sum of the two cantilever deflections.]

7. $k = 7.0 \times 10^6$ N/m. [Easily found from the static deflection due to the imposed mass.]

8. $m = 70$ kg. Wave height = 1.77 m. [This is the height above mean sea level; the peak-to-trough height will be double this value. The difference in apparent weight is due to the additional acceleration of the man's mass experiences due to the harmonic motion.]

9. $F(t) = 50 - \dfrac{400}{\pi^2}\left(\cos 2\pi t + \dfrac{1}{9}\cos 6\pi t + \dfrac{1}{25}\cos 10\pi t \right)$
[Treat original force as sum of a mean force of 50 plus a triangular wave centred on $F = 0$, then use result in Table C.1, Appendix C.]

CHAPTER 2

1. 4.36 Hz. [Use a standard beam deflection formula to find the vertical stiffness.]

2. $\omega_n = \sqrt{g/L}$ [By Archimedes' principle, the restoring force due to a deflection of the cylinder is equal to the weight of water displaced. Use the equilibrium case to eliminate the densities from the final answer.]

3. General case: $\ddot{\theta} + \dfrac{g}{R} \sin\theta = 0.$

   For small oscillations: $\omega_n = \sqrt{g/R}$ [Since $\sin\theta \approx \theta$ for small angles.]

4. $\ddot{\theta} + \dfrac{6k}{m}\theta = 0$ [Derive by equating moment of spring forces about pivot to moment of inertia × angular acceleration of bar.]

5. (a) $\omega_n = \sqrt{3k/2m}$, (b) $\omega_n = \sqrt{k/2m}$

6. $\xi = 0.607$ [Deduce by calculating the undamped natural frequency and comparing to the damped value given.]

7. (a) 47.7 s, (b) 9.5 s, (c) 2.4 s

8. $\xi = 0.71$, vibration amplitude reduces to within 1% of final value after 1.1 cycles.

9. Peak displacement = 8.7 mm, peak acceleration = 12.4 m/s². [Loading at 6 Hz is the worst case for both parameters.]

10. $k = 658$ kN/m. [Value has been calculated assuming system is undamped. Adding damping actually makes things slightly worse because it inhibits relative motion between the instrument and the floor, i.e. it tends to increase the vibration transmission.]

11. (a) Displacement = 177 mm, base shear = 420 kN, base moment = 5040 kNm.
    (b) Displacement = 48.5 mm, base shear = 115 kN, base moment = 1150 kNm.
    [Answers may differ slightly due to the accuracy with which you can read values off the response spectrum.]

CHAPTER 3

1. (b) $\omega_1 = \sqrt{k/m}, \omega_2 = \sqrt{3k/m}$, (c) $\mathbf{u}_1 = \dfrac{1}{\sqrt{2}}\begin{bmatrix} 1 \\ 1 \end{bmatrix}$, $\mathbf{u}_2 = \dfrac{1}{\sqrt{2}}\begin{bmatrix} 1 \\ -1 \end{bmatrix}$

2. [Following the normal procedure of applying Newton's second law to each mass gives the same equations of motion as in Problem 1. Hence, subsequent analysis steps will also be the same.]

3. $\omega_1 = \sqrt{g/L}$, $\omega_2 = \sqrt{g/L + 2k/m}$, (a) $k = 14.7$ N/m, (b) $k = 485.6$ N/m

4. [With no spring connecting the two masses, the damper provides the only coupling between the two degrees of freedom. For lightly damped structures, it is common to neglect the damping when determining natural frequencies – here, this will have the effect of reducing the structure to two unconnected single-degree-of-freedom (SDOF) systems, each with the same natural frequency.]

5. Equations of motion are (in units of kg, N, m):

$$10^3.\begin{bmatrix} 100 & 0 & 0 \\ 0 & 100 & 0 \\ 0 & 0 & 50 \end{bmatrix}\begin{bmatrix} \ddot{x}_1 \\ \ddot{x}_2 \\ \ddot{x}_3 \end{bmatrix} + 10^6.\begin{bmatrix} 100 & -50 & 0 \\ -50 & 90 & -40 \\ 0 & -40 & 40 \end{bmatrix}\begin{bmatrix} x_1 \\ x_2 \\ x_3 \end{bmatrix} = 0$$

6. $\mathbf{u}_1 = 0.00265\begin{bmatrix} 0.481 \\ 0.834 \\ 1 \end{bmatrix}$, $\mathbf{u}_2 = 0.00294\begin{bmatrix} -0.796 \\ -0.140 \\ 1 \end{bmatrix}$, $\mathbf{u}_3 = 0.00223\begin{bmatrix} -0.763 \\ 1 \\ -0.935 \end{bmatrix}$

[Remember that mode shapes can be normalised however you choose – for instance, just the term in brackets, in which the largest term is set to 1, would be an acceptable solution. The coefficients outside the brackets result in mass-normalised mode shapes.]

7. $a_0 = 0.897$, $a_1 = 0.00192$, $\xi_2 = 4.4\%$

8. Generalised forces in modes 1–3: $\mathbf{F} = \begin{bmatrix} 1978 \\ 584.5 \\ -505.3 \end{bmatrix} \sin 4\pi t$ (N),

Mode 1 displacement amplitudes at storeys 1–3: $\mathbf{x} = \begin{bmatrix} 0.086 \\ 0.149 \\ 0.179 \end{bmatrix}$ (m)

[This uses the mass-normalised mode shapes given in Problem 6. If you choose a different normalisation you will get different generalised forces, but should still get the same displacements – the normalising factor cancels in the end.]

CHAPTER 4

1. [The hardest part is getting the right boundary conditions at the tip; because a free end can support neither a moment nor a shear force, this implies the second and third differentials of the mode shape must be zero there. In fact, you only need one of these to solve the problem – the other would enable you to go on and solve for α, which you are not asked to do.]

2. [Answer for Problem 2 is given in the question.]

3. $x(z,t) = 16.8\sin\dfrac{\pi z}{L}\sin(8\pi t - 2.47)$(mm)

$M(z,t) = 58.9\sin\dfrac{\pi z}{L}\sin(8\pi t - 2.47)$(kNm)

[It is reasonable to neglect higher modes because loading is so close to resonance with fundamental mode, so this will tend to dominate. Also, since the loading is symmetric, the anti-symmetric modes will not be excited at all.]

4. $x(z,t) = 5.2\sin\dfrac{\pi z}{L}\sin(6\pi t - 0.075) + 0.0086\sin\dfrac{3\pi z}{L}\sin(6\pi t - 0.0034)$ (mm)

5. $m^* = \left(\dfrac{3}{2} - \dfrac{4}{\pi}\right)mL$, $k^* = \dfrac{\pi^4 EI}{32L^3}$, $f_n = 0.583\sqrt{\dfrac{EI}{mL^4}}$

[This compares quite well to the analytical solution in Table 4.1, in which the frequency coefficient is 0.560.]

6. (a) $f_n = 1.111\sqrt{\dfrac{EI}{mL^4(0.5+\alpha)}}$, (b) $f_n = 1.103\sqrt{\dfrac{EI}{mL^4(0.486+\alpha)}}$

[For the second mode shape, due to symmetry, we can evaluate m^* and k^* by integrating from 0 to $L/2$ and doubling the answer. The integral of mass × mode shape squared for the lumped mass is simply M. Natural frequencies by the two formulae agree to within 1% for all values of α between 0 (no lumped mass) and 100 (very large lumped mass). With no lumped mass, mode shape (a) is the exact solution. Mode shape (b) is based on the static deflected shape of a beam with a central point load, and so might be expected to give a better approximation of the true mode shape for large values of α.]

7. $\psi = 1.617\left(3.5\dfrac{z^2}{L^2} - 2.5\dfrac{z^3}{L^3} - \dfrac{z^4}{L^4}\right)$

[The term in brackets satisfies the boundary conditions, while the factor 1.617 scales the shape to a maximum value of 1.]

8. f_n = 7.9 Hz, x_{max} = 7.0 mm
[The accuracy can be assessed by comparing the natural frequency estimate with the analytical solution (calculated from Table 4.1) of 5.5 Hz. The SDOF estimate is therefore not very good in this case; this is because, although it satisfies the boundary conditions, the assumed mode shape is not a good match to the correct one, as can be easily seen by plotting them.]

9. (a) $p^* = \dfrac{wL}{\pi}\sin\omega t$, (b) $p^* = F\sin\omega t$, (c) $p^* = \dfrac{F\sqrt{3}}{2}\sin\omega t$, (d) $p^* = \dfrac{2}{\pi}mLX_g\omega^2\sin\omega t$

10. (a) 46 mm, (b) 73 mm, (c) 63 mm, (d) 164 mm (relative to the ground)

CHAPTER 5

1. Energy dissipated = 200 J per cycle. (a) c_{eq} = 16.2 kNs/m, (b) c_{eq} = 3.24 kNs/m. The force required to achieve the imposed displacement has a magnitude equal to the frictional slip force, 2 kN, and flips sign as the direction of motion changes. [This is because, with $\omega^2 = k/m$, the stiffness and inertia forces exactly cancel, so that the

applied load has only to overcome the friction force. Interestingly, the same force is required whatever the magnitude of the imposed harmonic displacement.]

2. $c = 762$ Ns$^{0.4}$/m$^{0.4}$, $\alpha = 0.4$, F_D(max) $= 913$ N. [Note that the units of c depend on the value of α.]

3. (a) 1.75 kJ, (b) 2.0 kJ. [There is a smaller energy loss in the first cycle because this includes an initial elastic deformation. In (a), the hysteresis loop starts at the origin and ends at (0, 5 kN). Subsequent loops both start and end at this point.]

4. Co-ordinates of the first three yield points are $(x, R) = (0.1$ m, 5 kN), $(0, -4.5$ kN), $(0, 4.5$ kN). [Probably the easiest approach is to visualise the bilinear spring as a combination of an elastic-perfectly plastic (EPP) spring with stiffness 45 kN/m and yield force 4.5 kN, and a linear spring with stiffness 5 kN/m.]

5. The SDOF system requires a time step of 44 ms for stability and around 3.5 ms for accuracy [taking one fortieth of the period as an approximate guide]. Choose the latter. The MDOF system requires a time step of 0.9 ms for stability. [Based on the period of the highest mode. It is hard to be sure whether this would be sufficient for accuracy, as it depends on the periods of the modes which make a significant contribution to the response. It would be necessary to experiment with different time steps.]

6. A: $F_b = 3.4$ kN, $S_d = 195$ mm
   B: $F_b = 16$ kN, $S_d = 97$ mm

CHAPTER 6

1. (a) $R_{xx} = \dfrac{1}{2}\cos\omega t$
   (b) The autocorrelation is a triangular waveform taking the value 0.5 at $\tau = 0$, reducing linearly to zero at $\tau = \pm 2$ s. This triangle then repeats periodically, with period 4 s. [Recall that the autocorrelation of a periodic signal must itself be periodic.]

2. (a) Mean $= 0$, mean square $= 0.5$, variance $= 0.5$
   (b) Mean $= 0.5$, mean square $= 0.5$, variance $= 0.25$

3. (a) $2\dfrac{\sin\omega}{\omega}$, (b) $2\dfrac{\sin\omega}{\omega}e^{-i\omega}$, (c) $-4i\dfrac{\sin^2\omega}{\omega}$
   [(b) and (c) can be deduced from (a) with the help of the time shift property given in Section 6.4.2.]

4. $Y(\omega) = \dfrac{1}{\pi}X(\omega) * T\dfrac{\sin(\omega T/2)}{\omega T/2}e^{-i\omega T/2}$
   [In the time domain, the sampled signal $y(t)$ is equal to $x(t)$ multiplied by a rectangular pulse like that in Figure P6.3(b); in the frequency domain, the corresponding relationship is a convolution of the Fourier transforms. The distortion to the underlying signal caused by the finite-duration sampling can be minimised by making the sampling duration T as large as possible.]

5. Peak displacement amplitude = 313 mm at t = 1.72 s (and at several subsequent points). Amplitude of free vibrations after end of loading = 270 mm.

6. (a) $H(\omega) = \dfrac{Y(\omega)}{\ddot{X}_g(\omega)} = \dfrac{-m}{-\omega^2 m + i\omega c + k} = \dfrac{-1}{-\omega^2 + 2i\xi\omega_n\omega + \omega_n^2}$

   (b) $H(\omega) = \dfrac{Y(\omega)}{X_g(\omega)} = \dfrac{m\omega^2}{-\omega^2 m + i\omega c + k} = \dfrac{\omega^2}{-\omega^2 + 2i\xi\omega_n\omega + \omega_n^2}$

   (c) $H(\omega) = \dfrac{X(\omega)}{X_g(\omega)} = \dfrac{i\omega c + k}{-\omega^2 m + i\omega c + k} = \dfrac{2i\xi\omega_n\omega + \omega_n^2}{-\omega^2 + 2i\xi\omega_n\omega + \omega_n^2}$

7. $H(\omega) = \dfrac{i\omega c + k}{i\omega c + 3k}$

8. $|H(\omega)| = \dfrac{1}{\sqrt{m^2\omega^4 + c^2\omega^2}}$, $\tan\phi = -\dfrac{c}{m\omega}$

9. Average power = F_0^2, $S_{xx}(\omega) = \dfrac{2F_0^2}{1+\omega^2}$

10. $S_{yy}(\omega) = \dfrac{2F_0^2}{m^2(\omega_n^2 - \omega^2)^2(1+\omega^2)}$, $S_{yy}(0) = 0.0033$ m²s/rad

11. Mean displacement = 0, variance = 0.00378 m². [Method is quite similar to Example 6.10, but requires a different transfer function because the input is a ground acceleration rather than a force.]

Further reading

References below are provided both as sources of more detailed guidance than is possible in this relatively slim text, and for general background interest. Web references were accurate at the time of publication but come with the usual health warning that I cannot guarantee they will be maintained indefinitely.

CHAPTER 1: INTRODUCTION

Tacoma Narrows

A superb, illustrated account of this infamous collapse, together with extensive links and bibliographies, is available at

http://www.lib.washington.edu/specialcoll/exhibits/tnb/

Earthquakes

Two very good general introductions to the subject are

Bolt, B.A. (1999) *Earthquakes*. WH Freeman, Basingstoke.
McGuire, B. (2006) *Global catastrophes: A very short introduction*. Oxford University Press, Oxford.

and a good web resource is: http://earthquake.usgs.gov/learn/

You can get a good feel for the effects of an earthquake on engineered structures and support systems by reading the reports of expert reconnaissance teams available at

http://peer.berkeley.edu/publications/earthquake_recon_reports.html
https://www.istructe.org/resources-centre/technical-topic-areas/eefit/eefit-reports

The Millennium Bridge

Key papers on the bridge's design, assessment and retrofit and on the lateral excitation mechanism:

Dallard, P., Fitzpatrick, A.J., Flint, A., Le Bourva, S., Low, A., Ridsdill Smith, R.M., and Willford, M. (2001) The London Millennium Footbridge. *The Structural Engineer*, 79(22), 17–33.
Macdonald, J.H.G. (2009) Lateral excitation of bridges by balancing pedestrians. *Proceedings of the Royal Society A*, 465(2104), 1055–1073.

Arup also maintains an extensive website on the bridge: http://www.arup.com/millenniumbridge/

Damping

Those looking for an understanding of damping that goes beyond the simple viscous dash-pot model used here should consult:

Bert, C.W. (1973) Material damping: An introductory review of mathematical models, measures and experimental techniques. *Journal of Sound and Vibration*, 29, 129–153.
Jeary, A.P. (1997) Damping in structures. *Wind Engineering and Industrial Aerodynamics*, 72, 345–355.
Smith, R., Merello, R., and Willford, M. (2009) Intrinsic and supplementary damping in tall buildings. *Proceedings of the Institution of Civil Engineers: Structures and Buildings*, 163, SB2, 111–118.

Some useful conceptual discussions of dynamics (as well as statics) can be found in

Ji, T., and Bell, A. (2008) *Seeing and touching structural concepts*. Taylor & Francis, Abingdon (Chapters 13–19).

and the accompanying website: http://www.mace.manchester.ac.uk/project/teaching/civil/structural concepts/home.htm

CHAPTERS 2 AND 3: SDOF AND MDOF SYSTEMS

These chapters cover the most widely used elements of structural dynamics. Several texts cover this material at greater length than here – among the best are

Chopra, A.K. (2011) *Dynamics of structures: Theory and applications to earthquake engineering*. 4th ed., Prentice Hall, Englewood Cliffs, NJ.
Clough, R.W., and Penzien, J. (1993) *Dynamics of structures*. 2nd ed., McGraw Hill, New York.

The earthquake spectra introduced in Chapter 2 are fully specified in the European earthquake engineering design code: EN1998-1. *Eurocode 8: Design of structures for earthquake resistance, Part 1: General rules, seismic actions and rules for buildings*. European Committee for Standardisation.

Human-structure interaction

Vibrations caused or experienced by humans in structures such as footbridges, stadiums and lightweight floors are considered in

Bachmann, H. et al. (1994) *Vibration problems in structures: Practical guidelines*. Birkhauser, Basel.
Ellis, B.R., and Ji, T. (1997) Human-structure interaction in vertical vibrations. *Proceedings of the Institution of Civil Engineers: Structures and Buildings*, 122, 1–9.
Institution of Structural Engineers. (2008) *Dynamic performance requirements for permanent grandstands subject to crowd action*. Institution of Structural Engineers, London.
Zivanovic, S., Pavic, A., and Reynolds, P. (2005) Vibration serviceability of footbridges under human-induced excitation: A literature review. *Journal of Sound and Vibration*, 279, 1–74.

Tuned mass dampers

The idea of using a secondary, oscillating mass as a vibration absorber was developed in the early 20th century and the pioneering work of den Hartog is still widely cited today. It is most readily available in recent reprints of his classic book:

den Hartog, J.P. (1985) *Mechanical vibrations*. 3rd ed., Dover Publications (Chapter 3).

An excellent, more recent treatment can be found in

Connor, J.J., and Laflamme, S. (2014) *Structural motion engineering*. Springer (Chapter 5).

CHAPTER 4: CONTINUOUS SYSTEMS

The following books offer a more thorough treatment of the dynamics of continuum systems, extending beyond beams to elements such as rods, wires, membranes and plates:

Reddy, J.N. (2006) *Theory and analysis of elastic plates and shells*. 2nd ed., CRC Press, Boca Raton, FL.
Weaver, W., Timoshenko, S.P., and Young, D.H. (1990) *Vibration problems in engineering*. 5th ed., Wiley (Chapter 5).

CHAPTER 5: NON-LINEAR DYNAMICS

A wide variety of time-stepping methods is described and their characteristics fully explored in the following texts. Of these, that by Petyt is probably the more accessible.

Petyt, M. (2015) *Introduction to finite element vibration analysis*. 2nd ed., Cambridge University Press (Chapters 12 and 13).
Zinkiewicz, O.C., and Taylor, R.L. (2000) *The finite element method, vol. 1: The basis*. 5th ed., Butterworth Heinemann, Oxford (Chapters 17 and 18).

Eurocode 8 (see full reference under Chapter 2 above) specifies permitted structural analysis methods, including time-stepping and ductility-modified spectral approaches. The issues surrounding seismic design and dynamic response of structures with permitted ductility are excellently discussed by:

Booth, E. (2014) *Earthquake design practice for buildings*. 3rd ed., ICE Publishing, London.
Elghazouli, A.Y. (editor) (2009) *Seismic design of buildings to Eurocode 8*. Taylor & Francis, Abingdon.
Fardis, M.N., Carvalho, E.C., Fajfar, P., and Pecker, A. (2015) *Seismic design of concrete buildings to Eurocode 8*. CRC Press, Boca Raton, FL.

CHAPTER 6: FOURIER ANALYSIS

While highly relevant to structural vibrations, the methods introduced here are more widely applied in other fields. Texts that focus on civil or mechanical aspects are therefore comparatively few, and probably the best is

Newland, D.E. (1993) *An introduction to random vibrations, spectral and wavelet analysis*. 3rd ed., Longman, Harlow.

The application of Fourier analysis methods to experimental vibration analysis is well covered by

Ewins, D.J. (2000) *Modal testing: Theory, practice and application*. Research Studies Press, Baldock, Hertfordshire.

Index

Milton Keynes UK
Ingram Content Group UK Ltd.
UKHW051949071024
449327UK00026B/2235